Heart of Darkness

Books in the *Science Essentials* series bring cutting-edge science to a general audience. The series provides the foundation for a better understanding of the scientific and technical advances changing our world. In each volume, a prominent scientist—chosen by an advisory board of National Academy of Sciences members—conveys in clear prose the fundamental knowledge underlying a rapidly evolving field of scientific endeavor.

The Great Brain Debate: Nature or Nurture, by John Dowling

Memory: The Key to Consciousness, by Richard F. Thompson and Stephen Madigan

The Faces of Terrorism: Social and Psychological Dimensions, by Neil J. Smelser

The Mystery of the Missing Antimatter, by Helen R. Quinn and Yossi Nir

The Long Thaw: How Humans Are Changing the Next 100,000 Years of Earth's Climate, by David Archer

The Medea Hypothesis: Is Life on Earth Ultimately Self-Destructive? by Peter Ward

How to Find a Habitable Planet, by James Kasting

The Little Book of String Theory, by Steven S. Gubser

Enhancing Evolution: The Ethical Case for Making Better People, by John Harris

Nature's Compass: The Mystery of Animal Navigation, by James L. Gould and Carol Grant Gould

Heart of Darkness: Unraveling the Mysteries of the Invisible Universe, by Jeremiah P. Ostriker and Simon Mitton

Heart of Darkness

Unraveling the Mysteries of the Invisible Universe

Jeremiah P. Ostriker and Simon Mitton

PRINCETON UNIVERSITY PRESS
Princeton and Oxford

Published by Princeton University Press, 41 William Street, Princeton, New Jersey 08540
In the United Kingdom: Princeton University Press, 6 Oxford Street, Woodstock, Oxfordshire OX20 1TW

press.princeton.edu

ISBN 978-0-691-13430-7

Library of Congress Control Number: 2012 950892

British Library Cataloging-in-Publication Data is available

This book has been composed in Minion with Ideal Sans

Printed on acid-free paper ∞

Printed in the United States of America

10 9 8 7 6 5 4 3 2 1

Contents

Acknowledgments

The unfolding story of modern cosmology, as it appears in the popular press, is often a simple linear parade of heroes whose achievements are presented as inevitable outcomes: Copernicus, Galileo, the Herschels, Einstein, Eddington, Hubble, Sandage, and then the Modern Paradigm. In fact, it is a more winding tale, with these leaders, in addition to their great contributions, having made serious errors and other players making essential contributions. The two authors, having been participants over roughly the last half century of this enterprise, and knowing rather closely a large fraction of the cast of characters, have wanted to add extra emphasis to some of the physicists and astrophysicists whose vital contributions have often been overlooked in the conventional story line. Specific examples are the Abbé Georges Lemaître, George Gamow, Fritz Zwicky, and Beatrice Tinsley. Numerous other living scientists have also been noted, but we have surely not been fair to the contributions of innumerable eminent scientists whose work has not even been mentioned, although their contributions to cosmology have been significant and sometimes critical. We did not set out to write a scholarly and comprehensive history of modern cosmology. Our weak apology is based on the necessity of choices to be made given the limitations of space in a book that only has attempted to present highlights of the tale, and those topics selectively picked from the large number of equally valuable strands that we might have emphasized. So, to our numerous colleagues whose contributions have been slighted or omitted, we offer our apologies; we know of and value your contributions, but rather arbitrarily chose to pick a small number of no longer present co-explorers whose work has been passed over too quickly in the standard accounts.

Both of us are, at the most fundamental level, hugely indebted to our numerous colleagues, in Princeton, in Cambridge, and throughout the world. Science is a cooperative, global enterprise and, of all modern sciences, astrophysics has perhaps the densest and richest branches of international interconnections. Thus any listing of those who have assisted and informed us will be painfully incomplete. Nevertheless, however partial and inadequate our remarks, certain individuals contributed so critically to our work that they must be acknowledged individually. At Princeton, Paul Steinhardt, Jim Peebles, and Jim Gunn have provided extraordinarily helpful historical background and scientific insight. They were, themselves, central players in this great enterprise and we are enormously indebted to them for their assistance in correcting errors, pointing out gaps, and generally providing wisdom. In Cambridge, Martin Rees and Donald Lynden-Bell have been, over our careers, steady fonts of insight and directional guides.

In the writing of this work Ostriker gratefully acknowledges the editorial assistance of his wife, the poet and essayist Alicia Ostriker, and his old friend, the editor Robert Strassler, as well as his editor from Princeton University Press, Ingrid Gnerlich. All three read draft after draft of the manuscript, providing innumerable vital suggestions, regarding both organization and verbal felicity. However strong or weak the final product, their generous and thoughtful assistance was essential in helping to transform the overly technical, literarily incoherent initial effort into its ultimate form.

Simon Mitton expresses his deep appreciation to his Cambridge colleague Michael Hoskin, the distinguished historian of astronomy and biographer of the Herschel family, with whom he has enjoyed a close friendship for forty-five years. Hardly a day passes without Michael giving Simon avuncular advice on how to be a plausible academic. Likewise Simon thanks Owen Gingerich, historian of astronomy at the *other* Cambridge, for much encouragement and guidance, given selflessly and warmly over several decades. Simon's wife Jacqueline Mitton, who is also a Princeton author, contributed many valuable suggestions on the development of the manuscript. He warmly acknowledges the research facilities support of the Master and Fellows of St Edmund's College, Cambridge, where he has appreciated the good advice of Michael Robson, Lee Macdonald, Bruce Elsmore, and Rodney Holder. It is also a great pleasure and privilege to thank his agent Sara Menguc and her colleagues for their support.

Preface

Cosmology Becomes Data-driven Science

Cosmology, our study of the nature, formation, and evolution of the universe, has been transformed in an extraordinary fashion since the two authors of this book were students in the 1960s. When we were doctoral students at Chicago (JPO) and Cambridge (SM), two powerful but competing models were in the air: the big bang and the steady state models of the universe. There were passionate advocates for each, and a scientist's views of the subject were considered to be a matter of *belief*. On almost a daily basis we were exposed to the strong opinions and arguments of the great minds that were battling to understand the universe. At any gathering of professional astronomers one might be asked, "Do you believe in the steady state theory?" or "What do you make of this big bang universe?" Popular writings about cosmology—then and to this day—reflect that early, almost theological, intellectual atmosphere. Cosmology rested precariously on a series of beliefs, because data and hard facts were so scarce.

In the past half century, cosmology has changed totally. It is now an information-driven precision science, thanks to spectacular progress in instrumentation and information technology. Of course, there are still big ideas, but those ideas are shaped and constrained by the flood of data from telescopes on Earth and in space. Observations have amply and thoroughly confirmed the big bang model as essentially correct. The Hubble Space Telescope and many other instruments have now given us an inventory and mapped the detailed geography of our local patch of the universe and, more dramatically, provided direct observations from further and further back in time and space, so that we are now able to consider the telescopes with which we peer at the cosmos as time machines. When the Hubble Space Telescope lets us study a piece of the sky distant from us by seven billion light-years, we are seeing the world as it was seven billion years ago, half the current age of the universe. Thus we can directly see and measure the differences between then and now, to chart the evolution of the cosmos. There is no need for speculation. Or, speaking more accurately, speculations over cosmic evolution can be checked by direct observations. Although we cannot see all the way back to the Big Bang, 13.7 billion years ago, we can directly chart the evolution of normal galaxies to almost the period of their birth pangs. Furthermore, orbiting radio telescopes allow us to see all the way back to the moment when photons finally emerged from the primal soup that had imprisoned them during their first 300,000 years since the big bang, thus giving us a view of the radiation that is the residue of that period. Hence, we can directly view and measure the tiny primordial fluctuations, which grew through the action of gravity to become the rich, nearby world of galaxies, stars, and planets.

In today's cosmological discussions, any theory *must* be consistent with the panoply of X-ray, ultraviolet, optical, infrared, and radio information accumulating in our databases, the observations that show us concretely what the universe is at our epoch, how it came to its present state, and how it started. Cosmological investigations are still not as grounded and verifiable as those in other disciplines, such as engineering, but they have largely lost that intoxicating whiff of natural theology. Just as our present knowledge of the geological and biological facts concerning our mother Earth has banished to science fiction the speculations over "monsters from the vasty deep," so also our previously untethered cosmic fantasies must now be constrained by the awesome and growing libraries of cosmological information.

Complementary to this grounding in fact, we have developed quantitative and testable theories based on the known laws of chemistry, physics, and mathematics, which provide the framework on which to hang these new observations. In principle, determining the growth of fluctuations due to gravity in the primordial matter from which stars and galaxies developed, based on Isaac Newton's well-tested laws of physics, is just as straightforward as calculating the path of a baseball hit into the grandstand or the motion of a ship in water. The calculations may be more complex, but they do not require mathematics or science about which we are uncertain. In a parallel development, the equipment needed to solve these equations is now available.

Computers, following Gordon Moore's famous law for the speedup of electronic chips, have increased in their power to do arithmetic by more than a factor of one million since the 1960s. Today we can code any given theory into large-scale computer simulations, start with the initial state as observed by our radio telescopes, crank the machines forward, using the physics of Isaac Newton, Albert Einstein, and Niels Bohr, and see if we do indeed reproduce, in the computer and the visualizations made from its output, the pictures of our locally observed world in all its rich detail. This process could work out correctly or it could fail. But there is no cheating possible. As the observations and the calculations become more and more accurate, there is less and less room for the hand-waving arguments about how things "must work" out in order to save the appearances.

We have discovered that we can actually follow this course both observationally (using our telescopes as time machines) and computationally; and we can chart the evolution of the universe with fair precision. The animations made from our computer simulations really do resemble the observed course of the universe's development as seen by using our cosmic time machine. However, our success is contingent. Our model of how the universe grew to what we see today can succeed only if we invoke the existence of two fantastic components that we call, for lack of better names, dark matter and dark energy. The discoveries of both of these entities were surprises, and there were at first many scientists who (understandably) resisted their introduction. They argued that we were simply adding extra wheels to a complex and inherently rickety mechanism to make things work. Worse, the initial proposals appeared to violate the modern scientific method, because there was no independent evidence for the presence of dark matter and dark energy. We have not yet found direct evidence for these substances in our earthbound laboratories. They are too dilute to detect easily on Earth (although many efforts to do so are under way) and only over the vast volumes of space does their presence make a real, observational difference.

However, evidence that dark matter and dark energy dominate the universe steadily piled up. Before long, a variety of independent lines of argument developed that forced astronomers to take dark matter and dark energy seriously. To clinch the argument, these several independent methods gradually converged on essentially the same values for the amounts of dark matter and of dark energy. Generally speaking in science, if the solution to each mystery requires a new substance with its own special properties, deep skepticism is justified. But the evidence has progressed in the opposite way, with each new observed phenomenon reconfirming our previous estimates for the amounts of dark matter and dark energy.

One example suffices to make the point. As we will see in chapter 6, dark matter was first found, in the 1930s, in giant clusters of galaxies, the largest self-gravitating entities in the universe. It was thought to reside in the space between the galaxies. Then in the 1970s it was found lurking in the outskirts of nearby normal galaxies, surrounding them like dark halos. When detailed calculations were made, the same cosmic abundance of dark matter could explain both of these observed phenomena and, more fundamentally the formation and evolution of both galaxies and clusters. In chapters 5 and 8, we will consider how all cosmic structure grew from tiny seed fluctuations under the influence of gravity to what we now find in the local universe. Gravity, as Newton showed in the seventeenth century, comes from concentrations of matter. In the 1990s, we discovered that the amount of matter and consequent gravity needed to cause the growth of structure was again "just right." The same amount of dark matter is required to explain the origin of structure as for the other two phenomena, the properties of clusters of galaxies, and the dark halos. Finally, we will discover in chapter 8 how our giant optical telescopes have recently found bright, distorted images of extremely distant objects, images that could only be understood as caused by the intervening lumps of matter acting as gravitational lenses, amplifying the image of the much more distant objects—an effect predicted by Einstein. Again, the amount of intervening matter needed to produce the images was just the amount needed to cause the other phenomena detailed above. Check, check, check.

Our modern cosmological edifice seems to have been constructed in a robust fashion—but, of course, only time will tell. We think that we have the picture essentially correct, but it would be a naive (and foolish) scientist who would assert that we are approaching the end of discovery, and that now, at last, "we have it right." When we have, at present, no strong clues as to the physical nature of either dark matter or dark energy, there is obviously still much to be learned. But, should we expect that there would be revolutionary discoveries that will contradict what seems now to be such a coherent story? Does scientific history go through jumps, when the paradigm changes, and all pictures are turned upside down? There is a school of thought that questions the validity of the normal scientific method and the concept of scientific progress. Its supporters find a compelling narrative in describing our changing views of the world as contingent and based more on social networking among investigators than on a real grasp of nature.

We think that a careful reading of the history proves this attitude to be incorrect. Throughout the long history of cosmology the principal thinkers typically did believe that they had the correct model, even as the model was changing. The fact is, since the emergence of modern science in the Renaissance, they usually were correct—but their views were also incomplete. Their observations and theories were founded on the "local" world available to them, with the bigger picture only unfolding as their horizons expanded. The journey that we will take is one of steady expansion as both our mental and observational horizons grew from our planet to our solar system to our galaxy to the expanding universe. And our temporal horizon has grown at a corresponding rate from the human, historical time of thousands of years to the several-billion-year history of the Earth to the possibly unbounded cosmic time scales. What we have discovered again and again is that our essentially correct picture of the local universe was embedded in a much larger cosmos, and that in this emerging world new and strange forces and materials were the dominant components, with the familiar constituents of the earlier model now seen as relatively minor local parts.

It would be wrong, of course, to overstate this evolution as a steady march of progress. In antiquity and the medieval period those who tried to understand the mechanism of the heavens would often resort to the adoption of ad hoc fixes to their models. The inclination to find a patch for any gaping hole in theory is always with us. Even Einstein made such a move when he introduced an arbitrary constant into his equations to make possible a static universe, in harmony with his preconceptions. But today, the flood of data from our ever multiplying observatories, examining the cosmos from above the Earth's obscuring atmosphere in ever greater detail, in an expanding array of wavelength bands, leaves little room for this type of error. Present-day scientists, driven by the facts as they find them, are confident that they have reached a plausible consensus view of the origin, the history, and the present state of the universe, that the modern paradigm, supported by so many disparate lines of evidence, does seem to be truly robust. But of course there will be new discoveries and new surprises ahead.

■ Outline of the Journey We Will Take

This book shows how humankind has reached its present state of understanding of the universe in which it lives. Although it is no longer intellectually fashionable to see the advance of our understanding as a course of inevitable progress, the earlier worldviews have not, typically, been proved wrong. Rather, as we noted earlier, they have been found to be incomplete and were incorporated into larger and more accurate pictures. In the prologue, we summarize the knowledge accumulated from the ancient world through the Renaissance and the early period of quantitative, observational science. Two thousand years ago the Greeks had a fairly accurate geometric model of the Earth-Moon-Sun system, they had discovered the precession of the equinoxes, and they had compiled the first star catalogs. The Copernican revolution, extended and enriched by the mathematical physics of Johannes Kepler, by Galileo's telescopes, and by Newton's universal law of gravitation, embedded that picture into a precise model for the solar system. During the eighteenth and nineteenth centuries, our solar system came to be known as part of a much larger disk of stars seen in the night sky and called the Milky Way. Surrounding this galaxy were the puzzling nebulae, and speculations raged as to whether these were gaseous phenomena in the outer parts of our own galaxy or alternate, island universes.

Chapter 1, "Einstein's Toolkit, and How to Use It," starts with the twentieth-century revolutions of relativity and quantum mechanics that produced the physical laws to be used in understanding the world around us. In chapter 2, The Realm of the Nebulae," our cosmic exploration commences, as telescopes in the dark skies of the new world became powerful enough to show Vesto Slipher, Edwin Hubble, and others that the mysterious spiral nebulae were part of an expanding system of galaxies, many of them similar to our own Milky Way. Chapter 3, "Let's Do Cosmology" and its more mathematical appendix (appendix 1), shows how we can understand some core physical ideas of cosmology, the mysteries of an expanding universe, with no more mathematics and physics than a good high school education would provide. In chapter 4, "Discovering the Big Bang," we put this world into the context of Einstein's equations and outline the modern synthesis of an expanding, evolving, originally very hot universe called the Big Bang. Discoveries made during the last half of the twentieth century, that the sky is filled with microwave (radio) background radiation and that the lighter chemical elements had been cooked in a cosmological furnace, confirmed this picture, and what has become the standard hot, big bang model of cosmology was accepted as verified by all who cared to study the matter.

Up until this point, theoretical investigations had focused on the evolution of the universe as a whole and on whether it would expand forever or ultimately stall and re-collapse. The actual objects in the universe, such as galaxies and the groups and clusters in which they are arrayed, were somehow taken for granted. In cosmology they were treated as just "there" and of unspecified origin. No one asked when and how these things, the observable building blocks of the universe, had been formed. But then, as we show in chapter 5, "The Origin of Structure in the Universe," finally, in the last quarter of the twentieth century, the modern synthesis for the origin of cosmic structure finally developed and with it ideas for the formation of galaxies and other large-scale, cosmological structures. This paralleled the growing realization that there were two fundamental, additional, rather strange components—dark matter and dark energy—whose nature was unknown but whose presence was essential to make the whole machine work.

The exciting discoveries of these two vital components at the heart of our universe, made in the last decades of the twentieth century, are detailed in chapter 6, "Dark Matter—or Fritz Zwicky's Greatest Invention" and in chapter 7, "Dark Energy, or Albert Einstein's Greatest Blunder." The gravitational forces arising from the dark matter drive the concentration of ordinary matter into galaxies. But, the ordinary chemical stuff that the planets and the stars are made of, the material that emits and absorbs light, is now known to be only some 4 percent of the whole— the icing on the cake. The cake itself is made of dark matter, dark energy, and electromagnetic radiation, with the dark energy apparently the yeast inflating the cake in an uncanny fashion.

This is the itinerary for the cosmic voyage on which we will take the reader in the following chapters. We summarize the journey, its conclusion, and the still open questions in chapter 8, "The Modern Paradigm and the Limits of Our Knowledge," and chapter 9, "The Frontier: Major Mysteries that Remain." It is exciting, it is new, and, dare we say, it is likely to be fundamentally correct. But it is by no means complete, since, as noted, we still have no idea what constitutes the dark matter and the dark energy. We embark on our voyage in the period of western history called classical antiquity, but quickly reach the Renaissance, when the wisdom of the ancients, preserved, refined, and transmitted by Islamic savants, began to filter into an intellectually backward but awakening western Europe. A growing reliance on three aspects of rational inquiry would transform not only astronomy but also all human inquiry into nature. The three key concepts were these: the application of direct measurement and observation, the introduction of mathematical modeling, and the requirement that hypotheses should be testable and verifiable.

Thus, the scientific method, as we now know it, was born during these Renaissance attacks on the astronomical model of scholastic philosophy. This new scientific method, whose test bed was the astronomical world around us, became the foundation upon which all future technological progress would be based, from electronics to the revolutions in biology. It has carried us to our current vision of the universe, as detailed in the final chapters, and there is no doubt that it will carry us further in the future.

Heart of Darkness

Prologue

From Myth to Reality

■ Astronomy: The Endless Frontier

When we look up at a clear, dark sky and are inspired with wonder and curiosity by the sight above us, we share a long and vibrant story with our ancestors—a quest to understand the nature, origin, and behavior of the glimmering points and patches of light above and around us. What are the heavens made of? And what is our own planet Earth's place within the cosmos that surrounds us? These questions occupied the philosophers of antiquity for centuries. They fascinated such luminaries as Nicolas Copernicus, Galileo Galilei, and Isaac Newton, and they continue to captivate the leading scientists of the present century. Ambitious surveys of millions of remote galaxies, as well as missions to map the energy released at the origin of the universe down to an exquisite level of detail, have since brought us tantalizingly close to understanding the nature, evolution, and fate of the universe. Yet, throughout the centuries since antiquity, we share the same source of inspiration: questions that first kindled the varied stories of mythology now inspire hypotheses driven by astronomical observation and our knowledge of the laboratory-based laws of physics. We also now understand that, in the 1945 words of Vannevar Bush, science is an "endless frontier": new discoveries will always lead to an expanded understanding and to still more questions.

Throughout the history of cosmology, natural philosophers and, later, scientists based their theories on the world they could see. They constructed the best hypotheses they could using our excellent human vision until their view could expand through improved technologies, like better telescopes and eventually space-based instruments of observation. Therefore, as we mentioned earlier, cosmology is a story of expansion—of vision, mind-set, and of the physical universe itself. As our ability to observe farther out into the cosmos developed, an expanding universe was revealed to us—and our understanding of the nature of the cosmos became both broader and more refined, while yet remaining incomplete.

The more we have discovered about the universe in which we find ourselves, the more clearly we have come to see that a darkness, a mystery, lies at its heart. While we have by now an extraordinarily good working model of the cosmos—so good that every prediction we make is validated by the subsequent, exquisitely precise measurements—yet the two most vital components of this model, dark matter and dark energy, remain shrouded from our understanding.

In this prologue, we will embark on the story of how humankind came to this modern understanding of the universe. This is a tale that, after a short excursion through the discoveries of antiquity, will lead us through the Renaissance and the birth of modern science and the scientific method, the Copernican revolution, Galileo's ground-breaking observational work, and Newton's foundational work on gravitation, up to the eighteenth and nineteenth centuries, in which we came to know that we are part of a congruence of stars called the Milky Way, and that our galaxy is just one in a vast sea of other such island universes (though this possibility was not confirmed till later). At that point in our narrative, we will find ourselves about to be swept into chapter 1 and the twentieth-century revolution that formed the basis for our modern paradigm of cosmology. However, before we rush headlong into the present, let us return to the beginning and consider the worldviews developed by other very thoughtful, inquiring minds of earlier times. The best way to comprehend our present view of the universe is through history, to see how, from a simple start, it developed over time, as observations and calculations were steadily assembled into the larger and more comprehensive picture that we have today

■ Charting and Modeling the Heavens

An hour or so after sunset, in the year 134 BCE, the astronomer Hipparchus (190–120 BCE) gazed at the emerging starlit night from his home on the island of Rhodes, and made an astonishing discovery. In the constellation Scorpius, he spied an extra star. No ancient watcher of the sky had ever before recorded the sudden appearance of a new star. Excited by this extraordinary event, he decided to compile an accurate catalog of the stars, perhaps thinking that it would be handy to have a checklist of star positions to refer to the next time a new star appeared from out of nowhere. In a burst of intense activity, from 134 BCE to about 127 BCE, Hipparchus spent long hours in his observatory, measuring angles. He used this information to compile a catalog of the positions of 850 stars. Hipparchus compared certain of his star positions with observations of about twenty stars made about 150 years earlier in Alexandria.

This led him to another startling discovery: the stars had moved eastward in position by about two degrees in 150 years. What this meant was that the entire celestial sphere (for the Greeks, the outer limits of the cosmos) was slowly moving. Hipparchus had discovered the precession of the equinoxes. Supposedly fixed reference points that lay at the heart of the cosmic coordinate system were slowly but steadily sliding eastward due to the slow precession of the Earth's axis by gravitational forces. By way of his careful observations, Hipparchus introduced vastly improved data into the geometrical models that were developed to explain the motions of the celestial bodies, and his elegant refinements lasted for three centuries.

Hipparchus was not the first in this tradition. Greek philosophers brought to the western world the belief that the natural world might be understood through measurements, mathematics, and reasoned argument. In the third century BC, Aristarchos of Samos had proposed a Sun-centered solar system with the planets in the correct order, and he had estimated the size and distance of the Moon and Sun from the Earth using valid geometrical arguments—but of course very approximate measurements. He realized that the proof of the Earth's annual motion around the Sun should show up in the slight apparent wobbling motion of the nearest stars ("parallax"), but it was too small to be measured with the naked eye.

The next young mathematician after Aristarchos to work on what we might term evidence-based astronomy was Eratosthenes. In the third century BC, he determined the size of the Earth using a geometrical method. He knew that on a midsummer's day the Sun is directly overhead at Aswan in Upper Egypt. In Alexandria, which is more or less due north of Aswan, he measured the size of the small midday shadow cast by a vertical stick (or gnomon). From the measured ratio of the length of the shadow to the length of the stick he used a geometrical argument to find that the distance from Alexandria to Aswan is 2 percent of the Earth's circumference. What is important here is not so much his accuracy, but his boldness in deciding that a property of the real universe could be established by combining geometry and measurements.

The Greek philosophers who had the greatest influence on later western thought until the Renaissance were Plato and his pupil Aristotle, best known for his contributions to political, moral, and aesthetic philosophy. Unlike the later, observationally motivated astronomer-mathematicians we have been discussing, Plato and Aristotle belonged to a legalistic and axiomatic tradition that did not rely on experiment and measurement. For example, Aristotle asserted that heavy bodies fall faster than light ones without mentioning tests of the claim. His brilliant rhetorical skills prevailed in the end, and his contributions to natural philosophy, while pernicious in retrospect, overshadowed the extraordinary methodolog-

Figure P.1. In the School of Athens there is a group of geometers looking at a slate. This reminds us that geometry had a central role in Greek cosmology. Claudius Ptolemy stands with his back to us, wearing a crown (he was often confused with the Ptolemies who ruled Egypt).

ical and observational contributions of the more pragmatic Greek astronomical investigators, who showed how, with only their eyes, their wits, and elementary geometry, they could determine the size of the Earth and Moon, the distance to the Moon and much else.

The Hubble Space Telescope and modern technology were not needed for the discoveries that the Greek astronomers made; every reader of this book has the equipment required to understand correctly our astronomical surroundings. And the methods they utilized, which were principally geometrical, were revived and extended by the Renaissance scientists who led in the true "rebirth" of natural philosophy.

By the peak of the Classical Era, the observational catalogs had become ever more detailed and accurate. Claudius Ptolemy (90–168 CE), a Roman citizen living in Alexandria, was a philosopher, geographer, astrologer, and astronomer who flourished 300 years after Hipparchus. Ptolemy's greatest astronomical work was published in about AD 150. We know it by its Latin-Arabic name, *Almagest.* It was the first attempt to produce a synthesis and analysis of all the useful astronomical knowledge then known to the ancients. Its authoritative status made it *the* textbook of astronomy for nearly one and a half millennia. (Copernicus is known to have used it.)

In the parts of *The Almagest* devoted to planetary motions, we see Ptolemy at his most innovative. He made two important adjustments to the model of planetary motion refined and used by Hipparchus. First, he allowed the Earth to be displaced a little from the geometrical center of the circular planetary orbits. Second, by a rather technical move, he arrived at an improved representation of the motion of Mars, which had long confounded (and would continue to puzzle) mathematicians.

In the centuries after Ptolemy, classical learning declined, then collapsed, and ceased to exist in Christian, western Europe. However, scholars in the Islamic world rescued the Greek texts from oblivion during the dark ages. Today we have reminders of the Islamic contributions to learning while Europe slept, through nouns such as algebra, algorithm, alkali, alcohol, zero, as well as star names such as Aldebaran and Algol, and so on. And then, eight centuries after Ptolemy, the spread of monasticism in western Europe led to the foundation of the first universities (Bologna 1088, Paris about 1150, Oxford 1167, Cambridge 1209) and the start of medieval learning, through which the masters and their pupils rediscovered ancient philosophy. In Arabic and European centers of learning, Ptolemy's *Almagest* was the standard work on planetary motion for about 1,400 years, until it was displaced by the Copernican revolution in thinking.

■ Copernicus: "The Last of the Greek Cosmologists"

Nicolas Copernicus (1473–1543) broke ranks with the medieval past by reviving the model of Aristarchos of Samos and postulating something quite incredible to those trained in the scholasticism of Aristotle and Thomas Aquinas: the idea that the Sun is the center of the planetary system. By 1514, the architect of the new solar system had made sufficient progress with his heliocentric theory that he felt confident enough to write a short essay, the *Commentariolus* (*Little Commentary*), which he circulated to certain astronomers. Copernicus claimed that his new model solved several of the problems of ancient astronomy. He had Earth as a center of gravity and the center of the lunar orbit. Otherwise, all planetary orbits, including the Earth's, encircled the Sun. The moving, rotating Earth itself created the apparent motion of the heavens. But the essay was no more than an extended letter, and Copernicus explained to the recipients that he was already writing a much larger work that would contain the full mathematical derivations, *De Revolutionibus Orbium Coelestium* (*On the Revolutions of the Heavenly Spheres*).

Copernicus continued to work on planetary motions for as long as two decades, slowly gathering more observations in an attempt to refine the orbital elements. However, while this theory lived in only a single manuscript, it could not affect cosmological thinking. But news of the daring scheme did percolate west. In Nuremberg, Georg Joachim Rheticus learned from a copy of *Commentariolus* that Copernicus had thrown the static Earth into seemingly violent motion. Rheticus decided that the radical cosmology of Copernicus needed further investigation. In 1539, he set out from southern Germany for the Baltic shores of northern Poland. Fortunately for Rheticus, the aging cleric welcomed the young enthusiast as a long-term guest.

Like a modern professor and a graduate student, the pair worked through the manuscript, taking many weeks to do so. We can imagine that at their initial academic sessions, Copernicus would have explained that several hypotheses were involved. The big idea was the cosmological device that put the Sun at the center of the solar system with the planets in orbit around it. As the details of the scheme unfolded day by day, Rheticus became convinced that the world should know what Copernicus had done. In 1542, Copernicus agreed that Rheticus could have a fair copy of the manu-

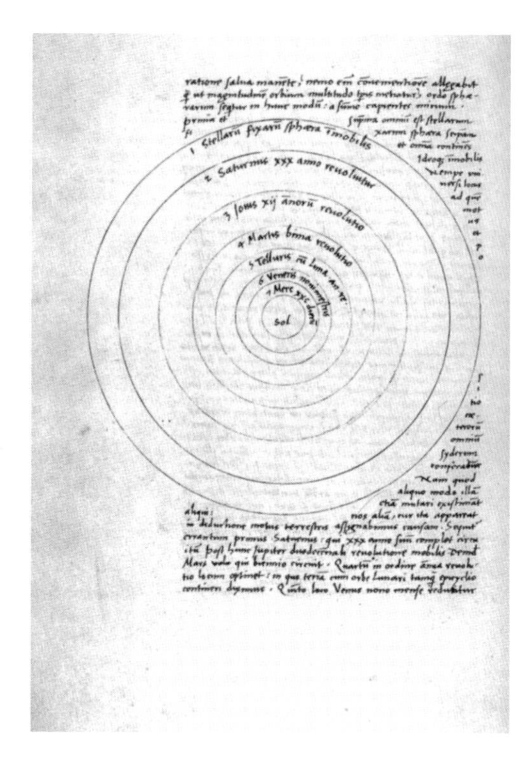

Figure P.2. The heliocentric model of the solar system in the duplicate manuscript that Copernicus kept. The circles are often described as planetary orbits. In fact, this is a two-dimensional representation of the nested spheres on which each planet's orbit is inscribed.

script to enable publication in Nuremberg. After spending many months at the right hand of Copernicus, Rheticus set out for Saxony, with the precious duplicate in hand. Finally, in the spring of 1543, the printing by Johannes Petreius in Nuremberg was complete. Hundreds of copies awaited distribution throughout Europe. On the title page the publisher's blurb boasted that the work is "outfitted with wonderful new and admirable hypotheses" from which the reader can "compute the positions of the planets for any time. Therefore buy, read, and profit."

The publication of this greatest scientific text of the sixteenth century, an epochal event, marks the dawn of modern scientific inquiry into the nature of the universe: its origin, its history, its architecture, and our place in the immensity of the cosmos. The scientific revolution that would transform Europe and then the rest of the world can be said to date from 1543, with the publication of *De Revolutionibus*. The revolutionary spirit that was spreading through the scholarly world is well summarized by the English scientist William Gilbert, an early advocate of the Copernican model, who wrote in 1600:

> In the discovery of hidden things and in the investigation of hidden causes, stronger reasons are obtained from sure experiments and demonstrated arguments than from probable conjectures and the opinions of philosophical speculators of the common sort.

The most persuasive advocate of the new approach and an extraordinary propagandist for what we would call the scientific method was Galileo Galilei.

■ Galileo: A New Approach to Mechanics and Cosmology

In recounting the events during the period from the birth of Copernicus (1473–1543) to the passing of Isaac Newton (1643–1727), we shall first turn to Galileo's (1564–1642) contributions to mechanics and cosmology. This pivotal figure in the development of modern physics and astronomy was pugnacious, brimming with sarcastic wit, properly respectful of authority in matters of church and state, but scathing with his scientific (or literary) adversaries. To this day he continues to be a controversial figure. In early life and middle age, he fizzed with energy regarding every aspect of physical sciences as then defined within the scope of his inquiries. He argued that science (establishing the term for the first time) must be based on measurement and on testable, mathematically formulated laws.

As is well known, Galileo laid the foundations for how objects move on the surface of the Earth (the science of mechanics). It was he who first realized that we should focus on acceleration, not velocity, and that forces on bodies cause their motion to accelerate, the increase in velocity being a by-product. Over a period of time, he devised a series of ingenious experiments to test the behavior of bodies in free fall and on sloped surfaces. It was through Galileo that the laws and language of dynamics developed. He tested his concepts of motion through the use of inclined planes. From these experiments he developed quantitative concepts of force, as a cause for accelerated motion.

Galileo understood that objects fall because a force, gravity, pulls them down, and of course he is most famous for establishing that all falling bodies, regardless of their weight, are accelerated by gravity at the same rate. While he may or may not have actually done the experiment from the Leaning Tower of Pisa, his conclusion was fundamental in encouraging Newton to propose that the gravitational force on all bodies was in proportion to their masses. He concluded that the natural state of an object experiencing no force is rest or uniform motion. Finally, in what some scholars have described as his greatest contribution to physics, he articulated the principle of inertia: a moving object will remain in its state of motion unless an external force acts on it.

Galileo must also be credited with persuasively advancing another fundamental tool of modern sciences: the mathematical method. The ancients considered natural law as a given, whereas Galileo showed that the laws of motion could be described using mathematics. He established a new and powerful mode of analysis that was followed by Newton and all successors in the physical sciences.

Galileo was not only an innovative mathematician, but also an observer. Many readers will have heard the oft-told account of Galileo's invention of the astronomical telescope and the discoveries made with it in 1609–10. Although spyglasses were already available in the Netherlands, it took the genius of Galileo to improve their optical performance sufficiently well for astronomical observation to be feasible.

By August 1609, Galileo had a telescope with 8x magnification, which he demonstrated to authorities in Venice, "to the infinite amazement of all." By early 1610 he had improved the magnification to 20x and then to 32x, at which point he started to look at the night sky. What did he see? Much is made, and rightly so, of his observations of Jupiter and its four moons collectively named in his honor, as well as his observation of the phases of Venus. At a stroke, these observations decisively demolished the Aristotelian cosmos; but they were perfectly consistent with the Copernican model. With his keen eye and brilliant mind, Galileo confirmed that the structure of the solar system was in accord with the Copernican heliocentric hypothesis.

Our interest lies in what Galileo discovered when he shifted his vision beyond the solar system, to the starry realms beyond. Wherever he aimed his telescope, he spied far more stars than can be seen by the naked eye. In 1610, Galileo released his findings in a popular book *Sidereus Nuncius*, or, *The Starry Messenger*. It had taken him just three months to complete, and it made Galileo a

SIDEREVS
NVNCIVS
MAGNA, LONGEQVE ADMIRABILIA
Spectacula pandens, suspiciendaque proponens
vnicuique, præsertim verò
PHILOSOPHIS, atque ASTRONOMIS, quæ à
GALILEO GALILEO
PATRITIO FLORENTINO
Patauini Gymnasij Publico Mathematico
PERSPICILLI
Nuper à se reperti beneficio sunt obseruata in LVNÆ FACIE, FIXIS INNVMERIS, LACTEO CIRCVLO, STELLIS NEBVLOSIS,
Apprime verò in
QVATVOR PLANETIS
Circa IOVIS Stellam dispatibus interuallis, atque periodis, celeritate mirabili circumuolutis; quos, nemini in hanc vsque diem cognitos, nouissimè Author depræhendit primus; atque
MEDICEA SIDERA
NVNCVPANDOS DECREVIT.

VENETIIS, Apud Thomam Baglionum. M DC X.
Superiorum Permissu, & Priuilegio.

Figure P.3. Title page of *Sidereus Nuncius*, the book in which Galileo describes his great telescopic discoveries of 1609–10.

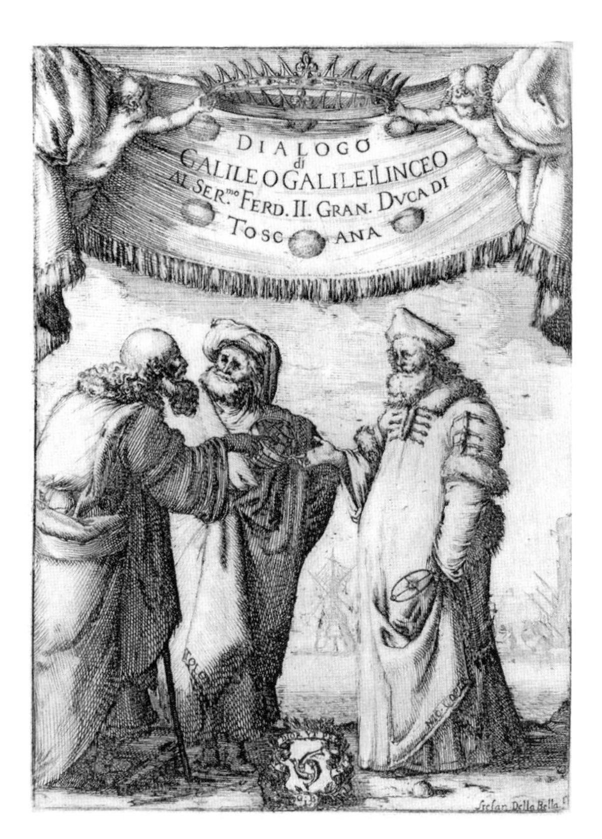

Figure P.4. Title page of Galileo's popular book on the geocentric and heliocentric models of the solar system. The engraving is a representation of Aristotle, Prolemy, and Copernicus with a maritime scene in the background. This was a popular book that got Galileo into serious trouble.

celebrity. In the constellation Orion alone, he quickly found five hundred new stars. He described how the telescope revealed "groups of small stars herded together in a wonderful way." Most spectacular of all, he resolved countless stars in the Milky Way, thus revealing for the first time that its structure is composed of discrete populations of stars:

> We have observed . . . the essence, namely the matter, of the Milky Way, which can be seen so clearly with the aid of the telescope that what for centuries philosophers found an excruciating problem has been solved with ocular certainty, thus freeing us from wordy disputes. For the Galaxy is nothing else than a collection of innumerable stars heaped together.

In 1611 he made a triumphant visit to Rome, where Cardinal Robert Bellarmine, head of the Roman College, asked his mathematicians for their opinion of Galileo's discoveries. They confirmed them all. From then on, until 1633, the infamous Galileo Affair clattered along in Rome, culminating in Galileo's trial for heresy and his conviction. Despite the continuing hostility of Catholic theologians to Galileo's work, by the mid-century the heliocentric model had pretty much swept the field, thanks in large measure to another of Galileo's popular science books, *Dialogue Concerning the Two Chief World Systems.* His international fame was such that the young English poet, John Milton, visited him during Galileo's confinement under house arrest in 1638.

■ The Impact of Copernicus: Kepler's Laws

Johannes Kepler, the German mathematician and astronomer, borrowed from Galileo the idea that one material object can invisibly exert a force on another. The concept that an invisible force field was at the heart of the solar system was to have extraordinary influence: as we shall see, it formed the basis for Newton's discoveries. Kepler imagined that the Sun was rotating and sending out an invisible influence that weakened with distance from the Sun. This leap of intuition led to Kepler's biggest breakthrough: the Sun controlled the planets through three laws. Expressed in modern form, the first law states that each planetary orbit is an ellipse, not a circle, with the Sun at one focus. At a stroke, one simple device—an ellipse—demolished all the perfect circles of antiquity and the medieval period. And, since the circles did not actually fit the orbits of the planets, the extra smaller and smaller epicycles, which had been added, could also be tossed aside at one stroke.

Kepler's second law states that the straight line drawn from a planet to the Sun traces out equal areas in equal times. When the planet is close to the Sun it moves more rapidly than when it is far from the Sun. This law swept aside the classical obsession with uniform motion (and in modern terms expresses the constant value of the angular momentum of a body in orbit). Finally the third law, the one that satisfied Kepler the most, is about heavenly harmony: the squares of the periods (the period being the time it takes for one rotation around the Sun—a year for the Earth) of planets are proportional to the cubes of their distances from the Sun. Through these laws, Kepler created celestial mechanics, in a final break with Aristotle. For the first time, an astronomer had demonstrated that the orbits in the solar system followed a common mathematical structure. Again, mathematics, specifically geometry, had triumphed in the heavens.

Working during much the same period as Galileo, the French philosopher René Descartes produced major works, but he was of a younger generation. A striking feature of Descartes' methodology is that it is driven by philosophy. He reveled in mathematics, becoming much impressed by the certainty that mathematics made possible, by way of its axioms, theorems, and proofs a precise description of reality. This led him to believe that geometrical reasoning could be used to probe the structure and mechanics of the universe. The cosmological structures that Descartes deduced from his metaphysics were extremely radical. The Cartesian model of the universe was packed solid with matter—there was no vacuum anywhere. Had there been no motion, there would be no difference from one part of the universe to another: the entire universe would have no structure. However, Descartes' philosophy supposed that there is structure in the cosmos, because matter is in motion; and, Descartes pictured matter in motion as vortices. The solar system was one such vortex, mostly made of invisible matter. In a period in which machinery, particularly astronomical clocks such as the one in Strasbourg Cathedral, offered a model of the universe, Descartes and his followers developed a mechanical philosophy. But—and the distinction is important—this philosophy was not science" in the terms spelled out by Galileo, as it did not involve measurements and testable hypotheses.

The next move forward came from Isaac Newton, who was born in the year of Galileo's death.

■ Isaac Newton and Gravity

Cartesian philosophy was all the rage when Isaac Newton arrived at Trinity College, Cambridge early in June 1661. The University was still officially in the thrall of Aristotle, as it had been since its foundation. Intellectual vigor had departed long before: learning was performed by rote, without enthusiasm. Cambridge was then a backwater: two-thirds of the students left without a degree, and the only relevant professional training was for a career in the Anglican Church. Newton found the official curriculum in a state of advanced decay. He became an autodidact: almost everything useful that Newton learned at Cambridge was the result of his own reading in solitude rather than from formal teaching.

Newton devoured the works of Galileo, Kepler, and others, and his student notebooks also reveal an interest in Descartes, though his investigations of light and vision eventually led him to reject Descartes' system of thought. He pointed out that, in the Cartesian system of vortices and pressure, eclipses would be impossible. He also made sketches, based on his own experiments, of the refraction of light through a prism. Crucially, he had the insight to realize that ordinary light is a mixture of colors. With that discovery, he understood that glass lenses would always produce colored halos around images. It was then a small step to invent the first reflecting telescope for astronomical research.

A second aspect of Newton's agenda interests us because he addresses the cosmic order and the nature of matter. He first taught himself Kepler's astronomy, after which he began to ponder the cause of gravity. What force exerted by the Sun could result in planetary orbits that satisfied Kepler's three laws with such precision? In December 1664, he sat up for many nights observing a comet, recording its changing position in his notebook. But then a natural phenomenon of a different order interrupted his research.

A disastrous plague had descended on England. When it reached Cambridge, the university was disbanded. At Trinity College the steward posted a notice at the Great Gate on August 7, 1665 stating that "All Fellows & Scholars which go now into the Country on the occasion of the Pestilence shall be allowed the usual Rates for their Commons in ye space of ye month following." In other words, everyone was sent packing with a month's wages. The university returned to normal only in the spring of 1667.

Newton, now with his BA degree, returned to Woolsthorpe to live with his wealthy mother. Newton's genius first flourished through private study during 1665–66. According to his biographer Richard Westfall: "The miracle lay in the incredible program of study undertaken in private and prosecuted alone by a young man who thereby . . . placed himself at the forefront of European mathematics and science."

It is worth recounting the steps Newton made toward his universal theory of gravitation. During the plague years he plunged deeply into mathematics, and developed the differential calculus that would be important to him because it gave him the tool needed for calculations on things that vary, such as the changing position of an orbiting planet with time. He also made advances in optics, obtaining the first good-quality spectrum of the Sun by projecting it with his prism onto a white surface and showing that it was composed of the colors of the rainbow.

What are we to make of the story of the apple? Numerous writers of popular science assert that Newton hit upon the law of universal gravitation as a flash of inspiration. In an early biography, his half-niece, Catherine Conduitt, said that the fall of an apple led young Newton to compare the gravitational force at the Earth's surface with that on the Moon. Yet, it is more likely that insight came more gradually. When the elderly Newton had reached celebrity status, he said: "I keep the subject constantly before me, and wait 'till the first dawnings open slowly, little by little, into a full and clear light." By 1666, he had the first dawning, but the bright light lay well in the future. Newton was elected a Fellow of the Royal Society in 1672. His new contemporaries in London were then hotly debating the mysterious influence that caused the planets to revolve around the Sun in ellipses. The sages asked: What is the mathematical form of the force law that stops planets from shooting off along a straight line into space?

In 1684, three famous men came to the brink of a mathematical solution to this question. The trio comprised the young Edmond Halley (of comet fame), Robert Hooke (Curator of Experiments at the Royal Society), and Sir Christopher Wren (a founder of the Royal Society, a former Savilian Professor of Astronomy at Oxford, and a highly esteemed architect). Halley already knew that for *circular* orbits the force of attraction from the Sun had to vary as the inverse square of the distance, but he was uncertain if the same law would apply to *elliptical* orbits. Wren declared his inability to solve the mathematical puzzle. Hooke, a brilliant scientist who continues to be underrated (perhaps because he was ugly and grumpy) had been carrying out excellent experiments on gravity for many years. In 1679, Hooke had written to Newton, saying that planetary orbits could be explained by "A direct straight line motion by the tangent and an attractive motion towards the central body." Hooke had lacked the mathematical skill to follow this up, and his letter invited Newton's opinion on the hypothesis. Newton replied that he had more or less given up natural philosophy for "other studies." The following year Hooke again wrote to Newton, this time stating explicitly that he supposed "the Attraction is always in duplicate proportion to the Distance from the Centre Reciprocall." That's the inverse square law. At the meeting of the Royal Society on January 14, 1684, Hooke claimed that all laws of celestial motion could be derived from an inverse square law.

Later in 1684, Halley quizzed Newton in Cambridge about "what he thought the Curve would be that would be described by the planets supposing the force of attraction towards the Sun to be reciprocal to the square of their distance from it." Newton's face brightened: he replied without hesitation that it would be an ellipse, at which Halley was "struck with joy and amazement." Two months later, Newton sent his friend a neat nine-page paper explaining the mathematical basis for Kepler's three laws.

An ecstatic Halley pored over that paper. In it, he found a grand synthesis: a single statement replaced Kepler's three empirically derived rules. Importantly, the nine-page paper hinted at an entirely new general science of dynamics, as well as an enormous advance in celestial mechanics. Edmond now hot-footed it back to Cambridge, where he persuaded Isaac to expand his short treatise. The end result was a full-length book that is now always known by its short title: the *Principia Mathematica*.

Newton immersed himself in the project. While he was writing *Principia*, Newton did nothing else for eighteen months. No alchemy. No theology. He forgot to eat. College servants attending to his needs complained that they saw "his Mess was untouched." Within weeks Newton had taken a ground-breaking step: he decided that *all* the celestial bodies attract each other. It wasn't just a case of the Sun attracting the planets, but rather the planets *also* attract each other. Eventually this step would lead him to the concept of universal gravitation: all clumps of matter, from atoms to clusters of galaxies, attract each other. In the Newtonian universe, all motion was subject to exact laws that applied throughout the entire universe, the structure of which would be influenced by the action of gravitational forces. Newton's path to the law of universal gravitation was a combination of precise observation with the mathematical formulation of a law of nature. It was to be an awesome synthesis, not to be improved upon until 1915.

As we have seen, Kepler's laws applied just to the planets, while Newton's synthesis gave a single force law that applied throughout the universe and could predict equally well the trajectory of an artillery shell or the orbit of a planet. Newton was not the sole originator of the laws of Newtonian physics, but he was the first to pull everything together into a logical framework that he titled "the mathematical principles of natural philosophy" which sets out three laws of motion, as well as the law of gravitation. His greatest contribution was to unify the laws of heaven and Earth into a single framework. And the method, the combination of precise observation with the mathematical formulation of a law of nature, became the model of science for centuries to come.

Figure P.5. Pierre-Simon Laplace (1749–1827), French pioneer of celestial mechanics whose work was translated into English by Mary Somerville. (Engraved by J. Pofselwhite and published in *The Gallery of Portraits: With Memoirs* encyclopedia, London: C. Knight, 1833. Shutterstock, Copyright Georgios Kollisdas)

While his three laws of motion worked exceedingly well for describing the motion of point particles, the mathematician Newton was presciently doubtful that they could apply to the universe as a whole. He was aware that, strictly speaking, the force of gravity could lead to instability; since the force is always attractive, the heavens might therefore collapse under their own weight. For that reason, the theologian Newton speculated that supernatural intervention would occasionally be required to prevent chaos in the cosmos.

In the eighteenth century, the impact of Newton's theory was most noticeable in France, where a succession of brilliant mathematicians applied themselves to analyzing motions in the solar system using Newton's laws of motion. Their soaring achievements did not in the end contribute to the quest to understand the origin of structure in the universe. However, we note in passing that, in 1796, Pierre-Simon Laplace (deservedly called the French Newton) published a popular account of the origin of the solar system. It is in complete conflict with the ideas that Newton had on this subject; he believed that the solar system had been created in its present form only a few thousand years earlier, and invoked a "God of the gaps" to explain the observed harmony and symmetry. In contrast, the same inexplicable harmony led Laplace to the concept that the system had arisen far in the past from a primitive rotating cloud, a "solar nebula." Laplace viewed the solar system as originating from the cooling and contraction of a flattened, slowly rotating cloud of incandescent gas. That picture survives to this day as a likely account of the nebular origin of the solar system. And on the largest scales, the contraction under gravity of great clouds of matter is today regarded as an important process for understanding why the universe has the structures we see and call spiral galaxies.

Now our journey will assume a different trajectory: we will pause in our exploration of theoretical developments in order to see how the remarkable astronomer William Herschel (1738–1822) discovered the stellar universe in the 1780s through observation.

■ William Herschel Discovers the Universe

In 1757, the musician and army bandsman William Herschel arrived in England as a penniless refugee from the Seven Years' War in which the French had ravaged his homeland, Hanover. In our narrative, the gripping story of how William picked himself up by his bootstraps and eventually established himself as an accomplished musician in the fashionable city of Bath Spa would be out of place. William enters our account in 1779: he has developed a keen, vocational interest in astronomy and is already accomplished at building superb, reflecting telescopes, which he uses to sweep the heavens. In this period, professional astronomers in London and Paris remained obsessed with applying mathematics to the solar system, neglecting the starry universe that was visible outside the planetary realm.

As an amateur astronomer, Herschel introduced a new methodology into astronomy: the concept of conducting a complete sur-

Figure P.6. *Left:* William Herschel, astronomer and musician. (Herschel Family Archives) *Right:* William Herschel's great 40-foot telescope (completed in 1789 for Observatory House, Slough, England) had a focal length of 12 m and a mirror with a diameter 1.2 m. It was the largest telescope in the world for more than half a century, being overtaken by the Rosse telescope in 1845. (Royal Astronomical Society Archives)

vey of all visible examples of a particular class of celestial body. One can think of this as an update of Ptolemy's catalog of a bit more than 1,000 stars, but with the enormous advantage that the telescope had replaced the human eye. Herschel's decision to survey the universe beyond the solar system is a key moment in the long history of attempts to understand the origin of structure in the cosmos (the largest current catalog, resulting from the Sloan Digital Sky Survey, contains data on over 260 million stars). He leaped ahead of the professionals by focusing on making new observations, rather than precisely measuring motions in the solar system.

In the cultured society of Bath, Herschel rubbed elbows with natural historians; and, following their example, he applied their methodology, of collecting specimens, classifying them, and then constructing a life history, to the realm of the stars. Before long, he decided to explore the entire universe on the same principles: observe the variety of celestial species, and arrange the specimens to create an evolutionary sequence. He was the first to apply this approach (natural history) to astronomy, a technique that the future Edwin Hubble would invoke in 1929 to classify galactic structure. When Herschel commenced in astronomy as an amateur, he read about a mechanical universe that ran like clockwork. He went on to breathe life into this machinery of cogs and wheels by showing that the milky nebulae had a life of their own, in which the force of gravity would cause nebulae to evolve, to become more compact with the passage of cosmic time.

Herschel swept the vault of the heavens with his trusty seven-foot telescope whenever there were clear skies. Then it happened. On March 13, 1781, the world's most accomplished amateur astronomer scoped a new planet during the course of his second survey. To cash in on the discovery, Herschel named the new planet *Georgium Sidus* in honor of King George III. The grateful king vested on Herschel a lifetime pension so that he could work full time on astronomy, and he later provided grants for new telescopes. We now know this "Georgian star" as Uranus, the seventh planet. Herschel became famous overnight. Renowned visitors descended on the Herschel household at 19 New King Street, Bath, anxious to meet the discoverer and examine his telescopes. Awards followed: in 1781, the Royal Society handed him the Copley Medal and elected him to Fellowship. The following year George III appointed the celebrity as his personal astronomer.

In 1783, Herschel began his epic work by sweeping his incredible 20-foot hand-made telescope across the heavens to record the distribution of both stars and nebulae across the entire, visible sky. The word "nebula" had been used since antiquity to describe celestial objects that present a blurred, misty, or milky appearance. Initially Herschel believed that all of the nebulae were just that: cloudy objects rendered misty by distance. He would later change his mind on that subject. At the outset, he knew that the survey would take many years; but, as he wrote to a friend, "it is to me far from laborious . . . it is attended with the utmost delight."

From this survey he attempted the first basic classification of the nebulae, according to their shape and structure, and with the explicit aim "to investigate the Construction of the Heavens." In the case of star clusters, he classified them according to their degree of clustering, ranging from "very compressed and rich" to "coarsely scattered." And, Herschel began to do what we would term astrophysics. He sensed that, within a cluster of stars, an attractive force is at work, and that the degree of clustering is a measure of the length of time over which the attractive forces have acted. Crucially, he realized that the variations he was seeing in the structure of nebulae and clusters indicated that some are older than others, and that they lie at different distances.

In a highly imaginative move, he began to investigate the structure of our galaxy. The general shape of the Milky Way—which is the insider's view of the galaxy—had been known since ancient times. But, to accurately delineate the structure of the Milky Way, Herschel invented a new observing technique that he named star-gauging. He simply counted how many stars he could see through the field of view of his 20-foot telescope, and then stepped across the sky, field by field, taking an inventory of the number of stars in each field of view. By 1785, he had made a systematic slice through the Milky Way galaxy, which he described as "a very extensive, branching, compound Congeries of many millions of stars," and sketches made in his notebooks illustrated the Milky Way's cross-section.

By 1790, he had come to the firm conclusion that nebulae were systems composed of stars. He felt that misty nebulae, such as Orion, were vast systems of stars at great distances. However, on the question of whether the universe was composed only of our Milky Way, or whether nebulae were starry systems, "island universes," outside our own galaxy, he changed his mind. The construction of the heavens beyond the solar system was, he thought, only made of stars, and nebulae were not actually other galaxies like our own.

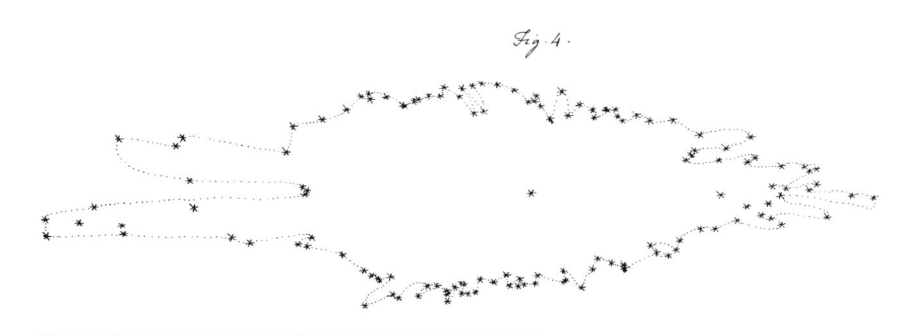

Figure P.7. One of William Herschel's cross-section sketches showing the structure of the Milky Way. This structure was deduced by the method of star counts. Herschel was of course not aware of the effects of interstellar dust, which obscured distant stars and unfortunately invalidated his method. (Royal Astronomical Society Archives)

Professional astronomers had essentially no interest in any of this, because none of them had telescopes that could compete with Herschel's. So the next step in resolving the puzzle of the nature of our universe, and whether it was composed only of the Milky Way, had to await the construction of even larger telescopes.

■ Understanding the Universe Becomes a New Kind of Science

By the middle of the nineteenth century, thanks to the work of Galileo, Kepler, Newton, and Herschel, in particular, the spirit of intellectual enquiry in Europe and the United States had turned decisively toward the goal of expanding humanity's vision of the cosmos and explaining how and why objects behave as they do in terms of mathematics and physics. In other words, the quests to understand the nature of stars, the meaning of the nebulae, and the overarching construction of the heavens were problems for which the solutions lay in classical physics and astrophysics, rather than positional and dynamical astronomy. In fact, the word and the scientific field of astrophysics was invented at the turn of the century when, in 1895, George Ellery Hale launched the *Astrophys-*

ical Journal to be the place where discoveries in the new science were to be published.

Among the most major problems that would occur to someone trained in physics was the one that any child might ask: why does the Sun shine? No known earthly fuel could provide enough energy, and if the energy was being provided by gravitational contraction, then a simple calculation made by physicists at the end of the nineteenth century showed that the Sun would only last for millions of years, much briefer than the age of the Earth, as known from the geological record. This obvious question and a host of others could be and were being asked for the first time as the laws of laboratory physics began to be applied to the universe surrounding us.

This was the intellectual *zeitgeist* leading up to the time in which our first chapter begins. A dramatic transformation was under way, in terms of the tools scientists would use to understand the cosmos and our place within it. Not everyone recognized this, and at the turn of the twentieth century, many thought that the laws of physics dominating the observable universe were so well-defined and complete that only a few, small refinements were still needed. The gaps in our understanding were overlooked. The majority of astronomers were also persuaded that there was little left to discover from an observational standpoint; the majority view was that the entire universe was composed only of our galaxy, the Milky Way, and that the well-understood solar system was to be found in the center of that disk-like assemblage of stars.

From our perspective in the current day, we can see that—in spite of the extraordinary leaps of thinking that occurred since we began this prologue—cutting-edge astronomy and physics at the turn of the century still had a long way to go. As it would turn out, the most astonishing advance in humanity's understanding of the laws of physics since Newton's proposal of a law of universal gravitation was yet to come. From the smallest structures, atoms, to the largest ones, galaxies and the universe, the new physical laws of quantum mechanics and relativity brought a revolutionary change to our understanding of the natural world. This is the subject of the next chapter. We will begin it immersed in the spirit of scientific hubris and complacency that characterized the turn of the century. While most scientists felt their toolkit of physical principles was essentially complete, a young physicist was about to show the world otherwise.

Chapter One

Einstein's Toolkit, and How to Use It

■ Overconfidence among the Cognoscenti at the Dawn of the Twentieth Century

As the nineteenth century drew to a close and the new century dawned, an intellectual ferment spread across the disciplines comprising western culture. Art, music, literature, and science were radically transformed in the Modernist period to a degree comparable to the changes that occurred in the Renaissance and the Enlightenment. The revolutionary expansion of our cosmic consciousness, which we will detail in the next chapter, was paralleled by the revolution in our scientific tools—the laws of physics. But at first no one saw the changes that were coming in both physics and astronomy. The complacent, turn-of-the-century belief that the laws of physics were essentially nailed down, with only refinements to follow, was blown apart by Albert Einstein, who singlehandedly initiated the revolutions of quantum mechanics, special relativity, and general relativity. In the next twenty years, he and his colleagues established the new physics that provided the foundations for our modern cosmology.

Let's begin our exploration of the physics revolution in 1894. Professors have gathered for a lecture in a new, neo-gothic building, a curious amalgam of the ancient and the modern at the just established University of Chicago. In his ceremonial, opening address, Albert Michelson, the director of the laboratory, said:

> The more important fundamental laws and facts of physical science have all been discovered, and these are now so firmly established that the possibility of their ever being supplanted in consequence of new discoveries is exceedingly remote. Future discoveries must be looked for in the sixth place of decimals.

How wrong he was. Michelson himself went on to become the most accomplished optical experimentalist of his age and to perform the Michelson-Morley experiment that helped to confirm Einstein's revolutionary new physics. Simon Newcomb, head of the United States Nautical Almanac Office, had dismissed the possibility of future astronomical discoveries somewhat earlier, in 1888, but then later gave us the best measurement of the speed of light.

Even today, books proclaiming the end of scientific discovery are published. As recently as 1996 one of us was asked on a popular television program to comment on a new book, *The End of Science*, which maintained that we had come to the end of discovery and in fact were at the end of the investigation of the big questions. It was fairly easy to make the case that we still did not know the answers to the really big questions that any child might ask: How did the world begin? How will it end? What is the world made of? Are we alone in the universe? Clearly, big questions remain to be answered.

But in the late nineteenth century, because Newton's laws of motion and gravitation seemed to have successfully explained dynamical phenomena in the heavens and on the Earth, it really did appear to many physicists and to astronomers that the operation of physical laws was clear and that our knowledge of the universe was essentially complete. The subject of astronomy was primarily understood to be measuring the positions of the planets on the sky, computing their expected positions, and comparing the two ap-

Figure 1.1. Albert Einstein in 1905, the "*annus mirabilis*."

as a clerk in the Swiss patent office in Bern. Here, in translation, is his summary of the breakthroughs, extracted from a letter he wrote in May 1905 to his friend Conrad Habicht:

> Why have you still not sent me your dissertation . . . you wretched man? I promise you four papers in return. The first deals with radiation and the energy properties of light and is very revolutionary . . . The second paper is a determination of the true sizes of atoms . . . The third proves that bodies on the order of magnitude 1/1000 mm, suspended in liquid, must already perform a random motion that is produced by thermal motion. The fourth paper is only a draft at this point, and is on electrodynamics of moving bodies that employs a modification of the theory of space and time.

The first paper, which is, according to some scholars, the most revolutionary in the history of physics, contains Einstein's suggestion that light comes not just in waves but in tiny packets of energy, *quanta of light*, later called photons. At this moment in history, Max Planck had already hinted that energy is "composed of a very definite number of equal finite packages," as he stated in December 1900 at the Berlin Physical Society. But well-established diffraction experiments—in which a beam of light, passing through two slits in an opaque wall, can be seen to interfere with itself—had demonstrated the *wave* nature of light.

At the heart of Einstein's first 1905 paper was the central question bedeviling physics: is the universe made of particles, or is it the unbroken continuum of electromagnetic and gravitational fields described by classical physics? Einstein effectively argued that light could be both. That is, the particulate nature of light is an intrinsic property of light itself rather than a description of how light interacts with matter. Henceforth physicists would adopt this duality: light could behave as either a continuous wave motion (classical physics) or as a stream of quanta (quantum physics). Einstein's work made necessary the developments, which are still not complete, that could unify the classical and quantum modes of describing nature. Other physicists, including Paul Dirac, did succeed in the next decades, uniting quantum mechanics with electricity and magnetism. However, a quantum theory of gravity continues to elude our grasp.

Einstein's second and third papers provided evidence, from already accepted experiments, for the reality of atoms and molecules. The theoretical physicist Max Born recalled in 1949, "At the time atoms and molecules were still far from being regarded as real." The three papers thoroughly demonstrated the exceptional creativity of young Einstein; but, important as these works of 1905 were, our interest lies primarily in Einstein's fourth paper on special relativity, which startled the world by proposing new properties for space and time, thereby introducing revolutionary new ideas about the nature of the universe to the world and forever altering our concepts of physical reality. The invention of special relativity successfully effected the merger of Newton's dynamics with electricity and magnetism.

But none of this touched on gravity, a central component of Newton's laws and at the heart of all cosmic investigations. Einstein realized the defect and remedied it a decade later with general relativity, providing, at last, the update for Newton that was consistent with all known macroscopic experiments and also consistent in its formulation with Maxwell's laws. However, as we absolute velocity, although observations can measure velocity with respect to some other observer: an experimentalist below decks in a smoothly traveling ship could never determine the ship's speed. Einstein's second postulate was that the speed of light through empty space is independent of the state of motion of the emitting body. At first Einstein had great difficulty in reconciling these two postulates with his thought experiments, which involved moving trains (a tradition that lives on in today's textbooks). His frequent visits to the huge Zurich train station with its synchronized clocks provided the setting. The solution really did come in a flash, and we'll use Einstein's thought experiment to explain it.

An observer is standing on the embankment of a railway track exactly halfway between two distant towns, A and B. This stationary observer sees two bolts of lightning *simultaneously* strike church steeples in towns A and B. At the same moment, there is a train on the tracks midway between A and B, and an observer inside the train is at the midway point. The train is traveling from A to B. The train observer is therefore moving toward B and away from A. While the light from B is rushing toward the observer on the train, the motion of the train will take the observer closer to the onrushing light signal from B and farther away from the light signal from A. The train observer therefore sees the flash from B *before* the flash from A.

This leads to a conclusion that would become immensely important in observational cosmology. Events that are simultaneous at one point of reference are not simultaneous for a moving observer at exactly the same spatial location. This may seem trivial, but it is not. It means that there is no absolute time. Time does not go tick-tock all over the universe independent of the observer's frame of reference. This shattered a premise that Newton had made in *Principia* more than two centuries earlier.

The equations of motion now became a little more complicated than the Newtonian versions. In order to write equations that would work perfectly in frames of reference that are moving at constant velocity with respect to each other, Einstein had to attach a fourth dimension, time, to the three spatial dimensions. That deft move altered the representation of geometry in the universe from three-dimensional graph paper to a four-dimensional space-time continuum. The future implications for cosmology were immense because, as we shall see, modern cosmology involves the interpretation of observations from parts of the universe that are both remote in time and are moving at enormous velocities relative to our vantage point in the Milky Way.

There is a fifth paper from the 1905 *annus mirabilis*. September finds Einstein penning another letter to Habicht in which he casually remarks that mass is a direct measure of the energy contained in a body, and he states that light carries mass with it. His paper, published in the journal *Annalen der Physik*, fills just three pages, but contains the most famous equation in physics, $E = mc^2$, imprinted on thousands of T-shirts to this day. This formula means that the energy, E, equivalent to a body of mass, m, can be found by multiplying it by c^2, the square of the speed of light. Because the velocity of light is so great, very small changes in the mass of an object can release or absorb enormous amounts of energy. The truth of this equation would be realized in atomic weapons, nuclear power, and as we will see, the shining of the Sun and most stars. It provided the essential clue to those scientists, half a century later, who tried and succeeded in solving the ancient problem of how the Sun continues to shine, of how it has sufficient fuel to continue, when all conventional sources of energy would have been exhausted long ago.

■ General Relativity

What did Einstein do next? He knew all along that the special theory was incomplete because it applied only to the restricted case of constant velocity motion and that the all-important effects of gravity had been neglected. What would be the situation for motion with *acceleration*? In 1907, a new thought experiment about what an observer in free-fall would experience suggested a way forward. Einstein's conceptual advance at this stage was the realization that the local effects of being in a gravitational field were indistinguishable from the effect of being accelerated. In other words, a person sitting in a chair in a windowless chamber is unable to tell whether the weight experienced, pulling them down to the chair, is a sign that the chamber is at rest in a gravitational field or that it is uniformly accelerating upward into outer space. No experiment that she could do in this room would enable her to distinguish between the two possibilities. This was the needed generalization of Galilean relativity.

The realization that gravity and acceleration were equivalent led Einstein to another thought experiment, which we will recast in modern terminology. Imagine that our windowless chamber is accelerating in outer space. A laser beam comes in through a pinhole on one wall. By the time the beam reaches the opposite wall it is a little closer to the floor because the chamber has accelerated in its motion. Consequently the trajectory is curved. The equivalence of gravity and acceleration then leads to the conclusion that light will appear to bend when it goes through a gravitational field. Einstein then calculated the amount of bending when a ray of light passes through the Sun's gravitational field. He concluded: "A ray of light going past the sun would undergo a deflection of 0.83 seconds of arc." It is very striking how Einstein once again deduced a testable effect by deduction from grand principles.

There was a very important astronomical by-product of this realization. If light from a distant star passing by one side of the Sun on its path to us would be bent in one direction, then light from the same star passing by the opposite side of the Sun would be bent in the opposite direction and it might occur that both beams would converge to be seen by an observer on Earth. Then we would see the same star on either side of the Sun—the Sun would be acting as a *gravitational lens*, doubling and magnifying the distant source. This effect, predicted by Einstein, was much later discovered and made to work as a practical astronomical tool; astronomers now commonly use intervening galaxies to image much more distant astronomical objects.

Einstein's conclusion that light would not travel in a straight line in the presence of a gravitational field meant that the geometry of space must itself be curved. Space would not have exactly the same

Figure 1.2. The galaxy cluster Abell 2218 in the center of this Hubble Space Telescope image lies 2 billion light-years away. The striking blue and orange arcs are highly distorted and magnified images of distant background galaxies, about 10 billion light-years away. The foreground cluster is acting as a gravitational lens, spreading out the background image into giant arcs in exactly the fashion predicted by Einstein. (NASA; ESA; J. Rigby, NASA Andrew Fruchter and the ERO Team; Sylvia Baggett, STScI; Richard Hook, ST-ECF; Zoltan Levay, STScI)

geometrical properties as the graph paper geometry of Euclid. Einstein could not, therefore, develop the theory by using the Euclidean geometry he had learned at school. More powerful mathematical techniques were called for. By 1912 Einstein, on his own admission, was in danger of going crazy in his search for a mathematical toolbox that could handle the properties of a gravitational field in curved space. In July, he consulted his friend Marcel Grossmann, a mathematician who had earned much better marks at Zurich Polytechnic than Einstein had managed.

Grossmann suggested that Einstein should try the non-Euclidean geometry devised by Bernard Riemann, who had extended the geometry of curved surfaces to include curves in three and even four dimensions. No thought experiment can help a human brain to envisage what a curved four-dimensional space looks like, but Riemann had discovered a method for defining the distance between points in four-dimensional space. Riemann needed tools more general than straight lines and vectors, and turned to "tensors" in order to keep track, in an orderly way, of the calculations. To properly describe a distance in curved space-time requires these complex entities—tensors—which, for the four-dimensional world that Einstein had postulated, had sixteen components each. When first proposed, the mathematics was considered so complex that the newspapers were fond of saying how few people in the world could understand the Riemannian description of general relativity, but today thousands of physics graduate students worldwide learn this material every year.

Einstein and Grossman began to collaborate. In 1913, they gave the first exposition of how Riemann's geometry in four dimensions could help in the search for a relativistic theory of gravity. Einstein struggled for almost another three years before he was satisfied. By November 1915, he had a full explanation for the advance of the perihelion of Mercury: his new general theory matched the observations so closely that he knew he must be right. Einstein corrected his earlier calculation of the deflection of starlight grazing the limb of the Sun: it amounted to 1.75 seconds of arc, twice the value computed from Newtonian physics. The general theory also predicted the gravitational redshift of light coming from the surface of a massive compact object. Photons, climbing out of the gravitational field of a massive object on their way to a distant observer, lose energy, and that corresponds to a change in their color, from blue toward red. And finally, rotating massive objects were predicted to slightly pull the space around them, an effect known as frame dragging. Although the frame dragging was anticipated for decades, the distortion effect on space-time due to the spin of the Earth was first observed directly in 2011 by the NASA satellite Gravity Probe B.

Einstein had arrived at a new theory of gravity, in which the equations described two processes. These are, first, how a gravitational field acts on matter, telling it how to move, and second, how matter generates gravitational fields in space-time, telling it how to curve. The equations themselves, in the full glory of their tensor notation, are on display in chapter 7, when we turn to the revival of interest in general relativity, as modern scientists searched for the missing ingredient needed to make cosmological models fit data from recent observations.

In November 1915, Einstein plunged into a four-week flurry of scientific activity, working "horrendously intensely" as he put it. On each of four successive Thursday afternoons he read a paper to fifty or so members of the Prussian Academy who assembled in the grand hall of the Prussian State Library in the heart of Berlin. Even as the lectures began, Einstein was still working furiously on the equations. His fourth lecture, delivered on November 25, 1915, described the field equations that capped his general theory of relativity. The immense intellectual effort left Einstein exhausted but elated. His first marriage had collapsed, war was ravaging much of Europe, and horrific slaughter was taking place on the Western front. It was not immediately clear what the impact of the new theory would be.

Although Einstein realized that the general theory of relativity could be used to construct models of the universe as a whole, he was not able to find any solutions to the equations. That was left to others. The equations are completely general. When motions are small with respect to the speed of light and gravity is weak, Einstein's equations become identical to Newton's laws. With relativistic corrections included, solutions to the equations allowed new predictions for motion in the solar system, as we shall see. And finally, when the global solutions were found, they helped in addressing fundamental cosmological issues of the fate of the universe, for example, whether it would expand forever or re-collapse catastrophically.

■ General Relativity Is Tested, Passes the Test, and Is a Sensation

While the European world was at war with itself, the Dutch astronomer Willem de Sitter located in neutral Holland acted as a middleman, by keeping astronomers in Britain and Germany in touch with one other, as direct communication was impossible. Einstein, by then a professor in Berlin, took care to send his latest papers on relativity to de Sitter, who then became the first scientist to work on their astronomical consequences. Within months of getting Einstein's ground-breaking work, de Sitter had drafted three papers that examined the implications of this new theory of general relativity from an astronomical viewpoint.

One day early in 1916, Arthur Eddington, the Plumian Professor of Astronomy at Cambridge University, sat in the southwest drawing room of the university's Observatory, opening his morning mail. His inbox brimmed with papers and correspondence submitted to the *Monthly Notices of the Royal Astronomical Society*. A package with Dutch stamps, postmarked Leiden, caught his eye. It was from his friend de Sitter.

The packet contained a copy, in German, of Einstein's paper setting forth the general theory, together with a paper by de Sitter, written in English. As Eddington read these papers, his pulse quickened: he immediately grasped the supreme consequences of its content. De Sitter's paper, entitled "Einstein's Theory of Gravitation, and its Astronomical Consequences; First Paper," gave Eddington a head start in understanding Einstein because its content included calculations on how the path of starlight must be altered by the gravitational field of the Sun. Two further papers submitted to *Monthly Notices* by de Sitter increased Eddington's interest.

He set himself the goal of testing Einstein's theory by observing the positions of stars close to the limb of the Sun at the forthcoming solar eclipse, which would take place on May 29, 1919, by which time he hoped the appalling great war would surely be over. It would be a clean test: during the full eclipse one would be able to see the background stars normally obscured by bright sunlight. There was a definite prediction that stars seen near the edge of the Sun would appear to be nudged very slightly from their normal positions in the heavens by the bending of the light rays from them caused by the Sun's immense gravitational field. That bending should be detectable by careful measurements. Since Einstein had predicted a degree of bending that was twice the amount derived from Newtonian physics, the astronomers could carry out an immediate test to see if this revolutionary new theory was correct or not.

When Eddington received these papers from de Sitter, Einstein was not well known in England and, in any case, the wartime situation encouraged the British scientists to denigrate their German counterparts. Eddington was an exception to this anti-German sentiment, partly because, as an observant Quaker, he took the pacifism promoted by his faith seriously. Einstein, too, was a pacifist, as Eddington knew.

The Plumian Professor of Astronomy at Cambridge embraced relativity enthusiastically, and set about promoting it, while at the same time maintaining a commitment to internationalism and his opposition to war. He was almost alone in protesting when British learned societies erased the names of German colleagues from their membership lists and refused to send journals to enemy countries.

Early in the war, Cambridge University had secured for Eddington an exemption from signing on for voluntary military service on the grounds that "one of the most important observatories in England should be safe guarded as far as national needs permit."

By this device Cambridge deftly avoided the public shame of having a protesting pacifist on the faculty. But this was not to last.

In reaction to shattering losses in France during the summer of 1916, the British government introduced compulsory enrollment through the Conscription Act that required men aged up to forty to fight. Eddington made a series of appeals against compulsory service to the military tribunal in Cambridge. The authorities were perhaps prepared to allow an exemption for scientific work, but Eddington nailed his colors to the mast, citing his pacifist belief as the reason for refusing to support the war by fighting or by working as a conscientious objector undertaking farm labor. Rather than putting his hands to the plow, Eddington planned to test one of the predictions of the general theory.

By 1918, the Cambridge Tribunal was steeling itself to imprison thirty-five-year-old Eddington, who seemed fully prepared to go to jail. But then his friend Frank Dyson, the Astronomer Royal, sought a personal exemption on the grounds that the Royal Society had already approached Eddington with a proposal that he should take the lead in testing Einstein's prediction of the gravitational deflection of starlight by organizing an expedition to observe the 1919 total eclipse. Sitting in the Guildhall in Cambridge, with Dyson's letter on the table, and with a gaunt Eddington in attendance, the tribunal grudgingly granted the exemption on condition that Eddington continued to prepare for the eclipse expedition.

On Saint Valentine's Day 1919, the Royal Astronomical Society Dining Club held a pre-eclipse dinner at which the guests included C. R. Davidson and Arthur Crommelin, who were bound for Sobral in Brazil, and E. T. Cottingham, who would accompany Eddington to Principe, a small island off the west coast of Africa. Early in March, the eclipse chasers sat in Dyson's study at Greenwich, the evening before sailing to Lisbon. They discussed the amount of deflection expected from Newton's theory, and the (revised) Einstein value, twice as great, which Eddington already fully expected to obtain. Cottingham asked, "What will it mean if we get double the Einstein deflection?" to which the Astronomer Royal replied in jest, "Then Eddington will go mad and you will have to come home alone!"

Cottingham and Eddington reached Principe on April 23. Eddington's notebook records, "We soon found we were in clover, everyone anxious to give every help we needed." On May 16, they secured test photographic plates, on which they honed their measuring skills. Eclipse day, May 29, dawned with a tremendous rainstorm. The rain eased by noon, and at 1:30 p.m. the astronomers got their first glimpse of a watery Sun, where the partial phase was well advanced. A race against time now commenced, as they prepared feverishly to carry out the photographic program in good faith, come rain or shine. At totality Eddington became so professionally absorbed in changing the plates that he did not see the eclipse. Working in the open, they did not have a spacious darkroom, and could develop only two plates each night, which Eddington then measured by day. On June 3, 1919, a day he never forgot, and later described as the greatest day of his life, Eddington measured the one plate that gave a result and it agreed with Einstein. Figure 1.3b shows one of the eclipse pictures with measured stars barely visible between horizontal lines.

Back in London, Dyson brilliantly stage-managed the official release of the results at a crowded joint meeting of the Royal Society and the Royal Astronomical Society on November 6, 1919. Sir Joseph Thomson, president of the Royal Society, set the tone by declaring that the eclipse result was "One of the greatest—perhaps the greatest—in the history of human thought." Sir Frank then took the floor, announcing that the measurements did not support Newton's theory of gravity. Instead, they agreed with the predictions of Einstein's new theory. In a final summary, Thomson pronounced:

> These are not isolated results that have been obtained. It is not the discovery of an outlying island, but of a whole continent of new scientific ideas of the greatest importance to some of the most fundamental questions connected with physics. It is the greatest discovery in gravitation since Newton.

Figure 1.3. Observations of the total eclipse of the Sun on May 29, 1919. Two stars whose positions in the sky have been moved slightly by the bending of light due to the Sun's gravity—as predicted by Einstein—are visible during the eclipse and are shown between horizontal lines above and to the right of the eclipsed Sun. Inset shows Sir Arthur Eddington, who organized the eclipse expedition to test Einstein's theory. (Royal Astronomical Society)

On hearing that, the London correspondent for the *New York Times* scurried to Fleet Street to send a special cable to his editor. The *Times* ran the story on the front page: "LIGHTS ALL ASKEW IN THE HEAVENS." Subsequent media coverage propelled Einstein to worldwide fame. Professionals in astronomy fully accepted that general relativity provided the right physical framework for exploring the universe at large.

Contrary to some versions of how science progresses, the acceptance of the new theory, after its experimental verification, was essentially instantaneous and worldwide. Less than five years had passed since the initial publication of the general theory of relativity. It was consistent with Newtonian theory within the limits in which the latter had been found over the centuries to be correct; and for the small but measurable effects where the two theories differed—with regard to the orbit of Mercury and the bending of light—the predictions of Einstein's theory were correct and those of Newton's were found to be false—in the jargon of the philosophers, they were "falsified." In addition, now there was a theory of motion and gravity that was intellectually consistent with Maxwell's theory for electricity and magnetism. Newtonian physics was not "overturned," but instead was seen as the correct treatment of gravitational dynamics for the common cases in which the velocities were small compared to the speed of light, and the gravitational potential was small compared to the square of the speed of light, c^2. Newton's physics was embedded in the larger vision of Einstein.

As we shall see in the next chapter, a comparably significant paradigm change was also occurring in astronomy at this time. Our position in the universe was changing from a picture in which the Sun sat at the center of the only galaxy in a static cosmos to one in which its place was in the outskirts of an ordinary galaxy, one of billions in an expanding, infinite universe. That change took longer to propagate through the scientific and popular imagination, but less than a decade was required. There was no organized resistance to either scientific revolution. It was unnecessary to wait for the proponents of the older viewpoints to die or be discredited.

The evidence was clear. It could be reproduced by any investigator, so acceptance was swift.

■ Cosmological Solutions to Einstein's Equations

Einstein's field equations, which embody the theory of general relativity, must govern the behavior of matter and energy in the four dimensions of space-time, and, in principle, they should apply to our entire universe or to any other universe. They are, however, complex and difficult to solve. The pioneers struggled to find solutions that corresponded to the real, actual, observed universe. As we have mentioned already, Willem de Sitter found the first solution that might conceivably apply to our own universe, but because the de Sitter world model was empty of both matter and radiation, it was obviously inadequate.

One thing that can be said for the de Sitter empty universe is that it does not suffer from the problems of instability in a Newtonian cosmology. This nasty problem, which was known to Newton, is simply stated. Gravity always pulls things together, and it is a long-range force, where distant masses can have a large effect. Unless gravity is balanced by the centrifugal forces of orbital rotation (as it is in the solar system), there is an ever-present danger of collapse. There is another related problem, that of "boundary conditions" that is, large amounts of very distant matter, of which we can have no practical knowledge, that can gravitationally dominate over the local forces, and cause immense and unpredictable motions. This realization, that the distant universe can have dominant local effects, was attributed by Einstein to the physicist Ernst Mach and subsequently termed Mach's Principle. A much quoted recent formulation of it by Stephen Hawking and Richard Ellis is the following: Local physical laws are determined by the large-scale structure of the universe. This means that cosmology must be understood, not as an entertaining afterthought, but as at the foundation of laboratory physics, which is an unsettling thought.

Einstein did find an ingenious solution to these problems. He proposed adding an additional, arbitrary, component to his field equations. This is the famous (or infamous) cosmological constant, the effect of which is to add a repulsive force that can balance gravity's attraction. We will discuss this further, as well as its modern incarnation, which we call dark energy, in chapter 7.

In 1917, Einstein published a static spherical model of the universe that apparently resolved the stability problem in Newton's universe. He had successfully constructed a universe that was neither expanding nor contracting on the large scale. He had two reasons for doing this. First, he thought that this would solve the problem of a universe subject to gravitational collapse that troubled Newton, since at all times and places gravity and the cosmological constant, the forces pulling matter together and those pushing matter, would balance each other. And, second, he thought that the static model was what the astronomical evidence required.

In fact, the evidence was already accumulating by this time that this goal was wrong: the universe was in fact expanding! In a place far from the intellectual capitals of the world, the desert town of Flagstaff, Arizona, the assistant director of the Lowell Observatory, Vesto Slipher, had already published (in January 1915) the news that the spiral nebulae seemed to be moving away from us at a considerable speed. It took another ten years for the data to accumulate before that view became mainstream. Figure 1.4 shows the recently graduated astronomer, Vesto Slipher (see chapter 2), and early redshift measurements of the "nebulae."

In March 1917, de Sitter found additional cosmological solutions to Einstein's equations, demonstrating that the general theory of relativity allows many possible models for the universe. When de Sitter added Einstein's extra repulsion term, the cosmological constant, he found that the universe sprang to life: space expanded. In retrospect this is not surprising, since these models contained too little matter to constrain any explosive tendencies in the universe. Much later the astronomical community looked back and saw the importance of this early work, but it was altogether too strange and theoretical to have very much impact at the time.

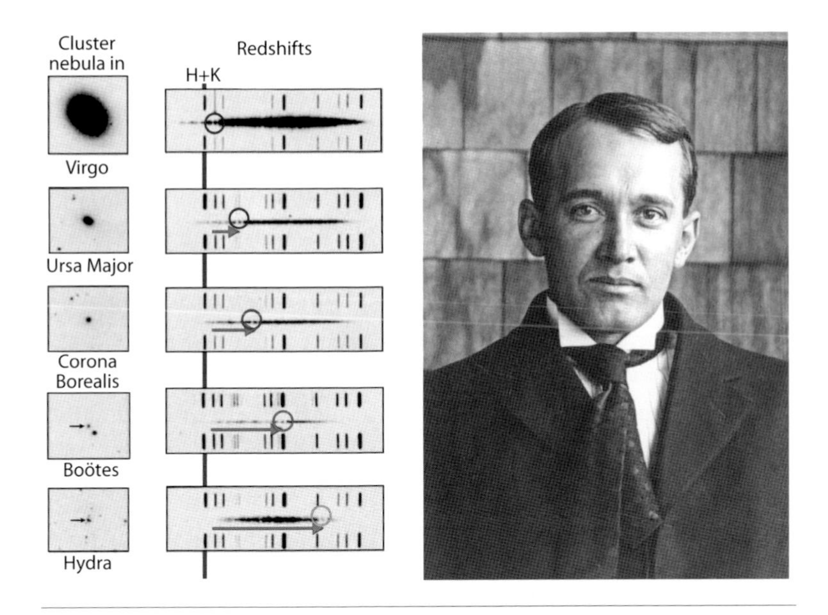

Figure 1.4. Vesto Slipher (*right*), who measured the spectra of many "nebulae," i.e., galaxies, and found (1915) that the lines in the spectra (*left*) were displaced by a considerable velocity toward the red (i.e., a "redshift"). His pioneering observations became the first evidence for the expanding universe. (Lowell Observatory)

De Sitter made his theoretical advances while the great European war raged on, and the hostilities had prevented progress in observational astronomy, except in the case of the United States, which suffered no physical harm. California had emerged as a center of excellence in observational astronomy. Mount Wilson Observatory, under the direction of George Ellery Hale, boasted a 60-inch telescope and the 100-inch telescope, then three years old and the largest in the world. American observers had begun to probe the contents of the universe beyond the Milky Way.

In Europe the obsession with relativity continued to grow once the war was over. In the 1920s, just two models of the universe were being considered: Einstein's matter-without-motion universe and de Sitter's motion-without-matter universe. Both were highly idealized and not based on any comparison with evidence from astronomy. Observational cosmology had not yet reached the stage where an informed choice could be made. Einstein was awarded the Nobel Prize in Physics in 1921, not (somewhat oddly) for general relativity, but for his discovery of the photoelectric effect, which, while an important contribution to the development of the quantum theory, was of much less import.

In a further unconventional twist, Einstein did not give the Nobel lecture at the presentation banquet in Stockholm in 1922. Instead, he gave the lecture in Gothenburg on July 11, 1923, with the title "Fundamental Ideas and Problems of the Theory of Relativity." He refers to his static model universe but clearly distances himself from it:

> It cannot, however, be concealed that to satisfy Mach's postulate in the manner referred to a term with no experimental basis whatsoever must be introduced into the field equations, which term logically is in no way determined by the other terms in the equations. For this reason this solution of the "cosmological problem" will not be completely satisfactory for the time being.

The cosmological constant, about which even Einstein was apparently dubious, survived, so to speak in the shadows, to be revived three quarters of a century later as the savior of big bang cosmology.

The next innovator in the saga initially shared de Sitter's fate of being overlooked: Alexander Friedman (also spelled Friedmann), a superb Russian mathematician, fought in World War I for imperial Russia. The Soviet Revolution of 1917, the civil war that followed it, and a subsequent blockade of the Soviet Union by the western powers, caused considerable delay before Soviet scientists became aware of general relativity. Friedman was the first mathematician in Russia to appreciate the importance of the theory for cosmology. In 1922, he wrote a classic paper on the curvature of space, in which he speculated that the appearance of the universe might change over time.

Friedman went on to show that the universe could have both matter and motion without violating Einstein's equations. Therefore, the universe could be expanding or contracting. Friedman taught at the University of Leningrad (now St. Petersburg), where the future cosmologist George Gamow attended his classes. His death in 1925 from typhoid fever had the unfortunate consequence that the significance of his work remained unrecognized for a further decade. The modern paradigm in cosmology directly follows from his work, and all of the standard variants still being discussed are culled from the Friedman models that he derived.

While much of Einstein's genius arose from the application of pure logic to well-conceived thought experiments, the overall scientific method was by now established. In order to earn widespread acceptance, any theory or model must be internally consistent and be formulated as a precise mathematical statement. It must also be testable by new experiments or observations; and it must be validated by those tests.

The mood of scientific complacency had been blown away and transformed into one of excited expectation. By the mid-1930s, the toolkit of physical principles was complete, but the tools had not yet been used to construct a viable world model. Chapter 2 provides the factual astronomical basis, the observational material, from which this world model is to be constructed, and in chapter 3, we describe the first attempts to use the toolkit for establishing a physically consistent cosmological model, one where the data will tell us directly whether the universe will ultimately re-collapse catastrophically or will expand forever as the stars fade away. T. S. Eliot, the iconic poet of modernism, had already (in 1925) opined on the matter: "*This is the way the world ends, Not with a bang but a whimper.*" Now it was time to see what the data would tell us.

Chapter Two

The Realm of the Nebulae

■ New Instruments in a Better Climate Unveil a New World

The 1919 British eclipse expedition confirmed Einstein's general theory of relativity, proving that theoretical astronomers had a toolkit they could use to investigate the properties of the universe. However, studying the remote cosmos soon became an activity in which the old world astronomers could not compete with those in the new world. Observatories under the clear and peaceful skies of Arizona and California emerged from wartime as the places where the discoveries were to be made. Mount Wilson Observatory, Pasadena, boasted a 60-inch telescope and the newly completed 100-inch telescope, the largest in the world.

American observers, using these new, more sensitive instruments, began to probe the contents of the universe beyond the Milky Way, where great discoveries awaited them. The world that they entered was not the old one of planets and stars, but an immense unexplored realm of galaxies and clusters of galaxies. At first, the nebulae being studied were of unknown nature. However, better and better observations soon showed conclusively that our own Milky Way is a disk galaxy in which the solar system has an off-center position and that most of the white, misty, nebulae were galaxies similar to our own. Moreover, toward the end of the 1920s, observations began to reveal that the whole of the cosmos is in a state of rapid expansion, with each galaxy moving away from all the others, and having a velocity proportional to the separations between them.

Using Einstein's toolkit, theoreticians could begin to find which of all the possible cosmological models actually fit the new data. In a few short years the Belgian scientist-priest Georges Lemaître found (in 1927) the solution that fitted the observations of the expanding universe codified by Edwin Hubble. This revolutionary change in the picture of the cosmos began in U.S. observatories at around 1910 and was essentially complete by 1930, a remarkably short period of time for such a dramatic change of our understanding of the universe to be formulated, promulgated, and then widely accepted. Although this overall picture of the extragalactic universe was widely adopted, the manifold implication of an initial explosive phase had not yet sunk in. The awareness of a fiery origin for the universe took decades more to be appreciated and will be treated in chapters 3 and 4.

In England, in the late eighteenth century, William and Caroline Herschel had made the first census of the nebulae and speculated that the attractive force of gravity could explain their appearance. But neither their work nor that of Lord Rosse in the 1840s (using his giant reflecting telescope) produced convincing evidence that the spiral nebulae lay beyond the Milky Way.

As late as 1890, the historian Agnes Clerke could airily dismiss as nonsense the idea that nebulae were "island universes." This concept had been introduced as speculation by the philosopher Immanuel Kant in 1755, but conclusive scientific evidence was absent and the idea remained on the fringes of respectability. The truth about the nature of the nebulae began slowly to emerge in the new world and the new century, at an unlikely location: a new observatory atop Mars Hill, at Flagstaff, in the Arizona Territory. With dark skies and few clouds this remote site was far superior to the cities in Europe that hosted the old observatories.

Figure 2.1. The 72-inch reflecting telescope erected by Lord Rosse in 1845 at the family seat at Birr Castle, Ireland. Following its restoration in 1997, it is the largest historical working scientific instrument in the world. (Jacqueline Mitton)

Percival Lowell, a wealthy Bostonian businessman, spent the last twenty-three years of his life devoted to astronomy. He moved to the clear mountain air of Flagstaff in 1894, where he used a 24-inch refractor telescope to explore the planet Mars. He thought he could see canals there and he really did believe what he saw. His 1908 book, *Mars as an Abode of Life*, proposed that Mars had once supported intelligent life, thereby kindling a new genre of science fiction writing. Despite his speculations and inclination to romance, Lowell was thoroughly scientific in his approach to astronomy. Although he hugely over-interpreted his own observations of Mars, his mission to understand the solar system was about to lead to a great discovery about the universe far beyond the solar system.

As a means of possibly finding out more about the formation of planetary systems, in 1901 Lowell had purchased a new spectrograph for his 24-inch telescope from the brilliant instrument maker John Brashear and hired a recent astronomy graduate, Vesto Slipher, to do the observing, while he attended to his many business interests in Boston. Slipher would receive missives from Boston, covering topics as mundane as looking after the vegetable garden ("when the squashes ripen send me one by express") as well as others directing the observing program.

Figure 2.2. Sketch of M51, made by Lord Rosse (William Parsons) in 1845 using the 72-inch telescope. This showed the spiral structure of a nebula for the first time. On top is a Hubble Space Telescope image of M51. (NASA)

Figure 2.3. Percival Lowell photographed at the Lowell Observatory, which he founded in 1894. (Lowell Observatory)

Slipher became an expert in the use of the spectroscope, studying the "spectral lines" drawn from the light emitted by stars and other astronomical objects. When light from any source is passed through a prism, it produces a spectrum, that is to say, it is broken into the rainbow of colors representing light at different spectral wavelengths. Within these color bands "lines" are sometimes visible, which give clues to the composition of the material producing the light. Each line from each chemical element is seen in the laboratory at a specific wavelength. But if the source of the light is moving with respect to the observer, then the wavelength recorded is shifted by a measurable amount, to longer wavelengths (redder) if the object is receding and to shorter wavelengths (bluer) if the object is approaching the observer. If the shift in wavelength of a specific line arising from a specific chemical element is measured (it is called the Doppler shift), it shows how fast a light source is moving with regard to the observer. An astronomical spectroscope measures velocities. Because Lowell's first priority was planetary research, he used the spectrograph to measure the rotation period of Mars, Jupiter, and Saturn.

Experimentation with different kinds of photographic plates and prisms led to significant improvement in the sensitivity of the spectrograph, which in turn led to an observational triumph: Slipher's spectra of stars in the constellation Scorpio revealed the existence of gas in the apparently empty space between the stars. This was a new and important discovery. It meant that at least some of the apparent nebulosity that had been seen was truly from fog-like gas, and this implied that the numerous astronomical nebulae could be classified as those that were truly gaseous nebulae, and those that were made up of densely packed stars. The latter appeared to be nebulous when looked at with telescopes that were unable to resolve them into individual stars and clusters.

Lowell offered Slipher encouragement and advice on probing the enigmatic spiral nebulae (which we now know to lie far outside our galaxy). A terse letter from Boston (February 8, 1909) directed Slipher thus:

> I would like to have you take with your red sensitive plates the spectrum of a white nebula, preferably one that has marked centres of condensation.

By "a white nebula" Lowell was referring to a spiral nebula. Lowell continued to believe that the spiral structures were planetary systems in the making and that the nebulosity was evidence for gaseous clouds similar to what Slipher had seen in Scorpio. Percival Lowell adopted Laplace's nebular hypothesis as the starting point for theories of the origin of the solar system and other planetary systems.

It took Slipher until December 1910 to get a spectrum of the great nebula in Andromeda, which appears as big as the Moon in Northern Hemisphere winters, its nuclear region just visible to the naked eye. It is now known to be a giant and somewhat older sibling galaxy of our Milky Way. Andromeda's nebulous appearance to our eyes is due to the simple fact that, at its great distance from us, its 100 billion stars blur together and cannot be seen as separate entities. Lowell, mystified that the spectrum showed faint peculiarities, tried to make the spectrograph more sensitive by removing two of its three prisms. This succeeded and squeezed the spectrum down to a thumbnail on the plate, but allowed much more light to reach the photographic emulsion. At this stage, Slipher in Arizona was engaged in a race against time because he had a competitor in California who was also wondering if the nebulae were a major component of the architecture of the cosmos.

William Campbell, director of the rival Lick Observatory, south of San Francisco, had proposed a program for making radial velocity measurements of the enigmatic nebulae to better understand their cosmic roles. The inspiration for that project had been the discovery by James Keeler, Campbell's predecessor at Lick, that great numbers of faint nebulae were recorded on long-exposure photographs. Thus the battle was joined: Campbell had the 36-inch refractor at Lick in California, while Slipher had to make do with the 24-inch at Lowell in Arizona. Neither party had any inkling of the enormity of the universe that was about to be revealed.

On September 17, 1912 Slipher's improved spectrograph made an astonishing discovery: the light from the Andromeda nebula were shifted to the blue end of the spectrum. The Andromeda nebula was approaching the Milky Way at considerable speed. Without question this was the first quantitative *extragalactic* observation, a measurement of the velocity of an object beyond the Milky Way. Slipher could not possibly have realized the magnitude of his achievement at the time. In fact, the discovery was so unexpected that he had misgivings. With a spectrum just a centimeter long and a millimeter wide, his technique did not really allow him to tease out the precious data encoded in the compressed image of the spectrum.

"I am of the opinion we can do much better," he wrote to Lowell, who responded immediately by shipping his own microscope from Boston to Flagstaff so that the wavelengths of the spectral lines could be measured on the tiny photograph of the spectrum with greater precision. With the microscope to hand, Slipher placed the fragile glass plate with its precious spectrum on to the stage of the microscope. He peered down the eyepiece tube, saw the shifted spectral line and confirmed that Andromeda had a significant blue shift; it was hurtling toward the Earth's home, the Milky Way galaxy.

For a result of this significance, confirmatory data was needed, and, at the very end of 1912, Vesto's great skill as an observer paid off. He exposed his final plate for three successive nights, finishing just as the New Year rolled in. During January he measured a total of four plates, spending almost three weeks meticulously reducing his data. His final result was absolutely astounding: Andromeda was racing along at 300 kilometers per second toward us! To put that into the context of velocity measurements at the time, the speed was about ten times higher than the average for stars in our galaxy.

Slipher's competitors at the Lick Observatory had a stinging response: "your high velocity is surprising in the extreme, the error must be pretty large, I hope you have more than one result." Ironically though, it would fall to Lick to confirm Slipher's triumph.

Lowell sent his congratulations and coaxed Slipher into observing more spirals. Slipher's spectrum of the Sombrero galaxy in Virgo (NGC 4594) brought great relief: he clocked it speeding at 1,000 kilometers per second *away* from Earth. That clinched it: the spectral shifts must be due to motion, and the motions were far larger than any previously found for heavenly bodies. Thus encouraged, he slowly accumulated data on other spiral nebulae. His count reached fourteen by the summer of 1914, and a trend was emerging: mostly the nebulae were receding rather than approaching us—Andromeda being the notable exception. Figure 1.4 shows Slipher and some early redshifts of "nebulae."

How did other astronomers view these results? Percival Lowell's outlandish claims about life on Mars had damaged the reputation of the Flagstaff observatory. Slipher possibly felt that his boss was casting a long shadow. However, a letter to Slipher from the Danish astronomer Ejnar Hertzsprung came straight to the point: the radial velocity measurements meant that the spirals could not belong to the Milky Way galaxy.

The velocities were so great that the gravitational field of the Milky Way could not hold them. In August 1914, Slipher presented his findings at Northwestern University in Evanston, Illinois, at the seventeenth meeting of the American Astronomical Society. Young Edwin Hubble, a graduate student at Yerkes Observatory (where one of us, Ostriker, would later study), was in the audience.

Slipher's amazing news led to a standing ovation, and even Campbell of the Lick Observatory congratulated him.

By 1917, Slipher had spectra showing measurable Doppler shifts for twenty-five spiral galaxies. His data revealed that three small systems and Andromeda (all of them relatively nearby objects) were approaching the Milky Way, and twenty-one more distant ones were receding from the Milky Way. The highest velocity that he measured was 1100 km per second. Slipher explained the velocities as being due to the motion of the Milky Way through the universe of nebulae. He was convinced by 1917 that the spiral nebulae were Kant's island universes, and that the Milky Way itself was a spiral nebula viewed from within.

While Slipher worked on spectra at Flagstaff, Heber D. Curtis at the Lick Observatory had been engaged for nearly a decade on photography of spirals using the 36-inch Crossley refractor. He too became an enthusiastic supporter of the concept that spiral nebulae are galaxies in their own right and independent of the Milky Way. As we noted earlier, there was another, major rival observatory, the Mount Wilson Observatory located in the San Gabriel Mountains of southern California, founded by George Ellery Hale in 1904 with funds from the Carnegie Institution of Washington. There, a 60-inch reflector was completed in 1907 and in 1917, the world's largest telescope, the Hooker 100-inch reflector was finished. Hale realized the enormous significance of the new results and, after hearing a lecture by Curtis in March 1919 on spiral nebulae, proposed a collaboration during a dinner in Washington. A few months later, Hale wrote to the director of Lick Observatory formalizing the proposal of a new collaboration:

> [At Mount Wilson] we are planning an extensive attack on spirals, with special reference to internal motion, proper motion, spectra of various regions, novae, etc., and here again I should be glad to know what Curtis has in hand so that our work may fit in with it to advantage.

The "extensive attack on the spirals" then became an industrial scale campaign. Even at this distance in time, one can feel the building excitement: these astonishing new observations being made by the new telescopes on the West Coast were surely portending a major scientific advance. But assimilating the evidence was to take some time.

The National Academy of Sciences (NAS) in Washington, DC, founded by Abraham Lincoln, was soon to convene a great debate on the implications of these discoveries. William Hale, the father of George, was a generous philanthropist who had endowed a fund for the National Academy of Sciences to be used for invited lectures. Not surprisingly, his son George had a considerable say over how these paternal funds were spent. Thus the program for the William Ellery Hale Lecture of April 1920 was chosen (after some discussion, wherein general relativity was rejected as the topic) to take the form of a debate concerning the new observational measurements of the spiral nebulae. G. E. Hale set up the debate between Heber Curtis of Lick and Harlow Shapley of Mount Wilson to take place on February 18, 1920, offering each a $150 honorarium for a giving 45-minute presentation on the subject, "the distance scale of the universe." The real, hotly contested topic was the interpretation of the new observational results: were the spiral nebulae gaseous objects in our Milky Way, or were they distant island universes?

Harlow Shapley, educated at Princeton under Henry Norris Russell, was Hale's junior associate at Mount Wilson. He was then thirty-four years old and had for several years been working on the extent of the Milky Way galaxy, using observations of variable stars (stars whose brightness oscillates in a regular way) as well as globular star clusters (round stellar systems containing roughly 100,000 stars each) to gauge the distance scales of the galaxy and its contents. He had become convinced (and he was correct in this) that the center of our Milky Way galaxy was nowhere near the Sun. He also realized that our galaxy was much bigger than had previously been thought, obtaining a distance to the center of our galaxy of 8.7 kpc, not far from the modern best estimate. He used the apparent brightness of the globular clusters to determine the distance scale of the galaxy, and he used the fact that they seemed to be ar-

Figure 2.4. Harlow Shapley. His 1939 paper reported that Cepheid stars in the center of our galaxy were 8.8 kpc from us, greatly enlarging the measured size of our Milky Way galaxy. (AIP Emilio Segre Visual Archives, Shapley Collection)

ranged in a spherical distribution to establish the location of the center of our galaxy. From our vantage point it lies in the zodiacal constellation of Sagittarius.

The changes made by Shapley in our conception of the world around us paralleled those initiated by Copernicus. The solar system was demoted from a privileged central position in the Milky Way to a spot now known to be about 28,000 light-years to one side. We live near a very ordinary star, orbiting with billions of others far off center. Shapley's picture, with relatively minor later modifications, has been amply verified by later work. But, somewhat oddly, during the famous debate he clung to the belief that our Milky Way occupied a special, central place in the cosmos.

That debate took place on the evening of April 26, 1920 in the Baird auditorium of the Smithsonian Museum of Natural History. Shapley, who lectured badly, read his address from a nineteen-page typescript. His approach was so elementary that it was completely uncontroversial: it took him six pages just to reach the definition of a light-year. Curtis, an accomplished public speaker, had his talk summarized on typewritten slides. In their verbal tussle, the pair wrangled over the distance scale of the universe, the debate's title. Curtis argued that the universe is composed of many galaxies like our own, which had been identified by the astronomers of his time as spiral nebulae.

By contrast, Shapley maintained that spiral nebulae were nearby gas clouds, and that the universe was composed of only one gigantic galaxy, our own Milky Way. As we noted, Shapley placed our Sun far from the center of a large galaxy and Curtis placed our Sun near the center of a relatively small galaxy, one of many in the universe. Each was correct on some points and incorrect on others. Scoring points in debates is not a very good way of establishing truth in science, and this event was no exception to the general rule.

The immediate reaction to the debate was influenced by the two men's different styles of public speaking. Writers of astronomical textbooks incorporated the debate into their books after 1921 (the year in which Shapley and Curtis published extended papers based on their lectures) and tended to give the nod to Shapley. To witnesses, the sparkling rhetoric of lively Curtis was effective, but dull Shapley may have won on substance.

By 1924, Curtis was proved correct about the nature of the spiral nebulae, and Shapley correct on the geometry of the Milky Way. But thanks to Edwin Hubble we now know that the observable universe is composed of billions of galaxies. The spirals are indeed galaxies just like our own.

■ A Universe of Galaxies Is Confirmed

By the late 1920s Edwin Hubble, whose early work on galaxies incorporated Slipher's data, had obtained sufficient new data from the 100-inch telescope to establish that the universe is expanding. The galaxies, with few exceptions, were apparently fleeing from one another. What sprang from this observation was the breathtaking concept that the universe had ballooned out from something much smaller at earlier times. Hubble's contributions to opening up the universe beyond the Milky Way galaxy can scarcely be exaggerated. But his work progressed in stages, with the first stage being to establish that we live in a universe of galaxies stretching out in all directions, with most of them superficially similar to our own Milky Way.

Ever a careful and meticulous observer, Hubble had use of the largest telescope in the world. He had arrived at Mount Wilson in 1919, the first year of operation of the 100-inch. After studying mathematics and astronomy at Chicago, he spent three years at The Queen's College, Oxford where he read law and Spanish as one of Oxford's first Rhodes scholars. At Oxford, he acquired some of the old-fashioned mannerisms, pipe smoking, and dress style associated with the English upper classes. He wore tweed jackets, knickers (plus-fours), and unnecessarily carried a cane, which irritated some of his astronomer colleagues.

At the outset of his career at Mount Wilson, Hubble first turned his attention to the spiral nebulae in Andromeda and Triangulum (M31 and M33). The mighty 100-inch telescope required many hours of tedious manual guiding. Fortunately for Hubble, the observatory director, Hale, had promoted Milton Humason, a mule-driver and janitor who had finished his formal education at the age of 14 (the norm at the time), to be Hale's night assistant, responsible for carefully aiming and correcting the giant telescope for hours at a time. Given his lack of high school education, Humason's progress as a professional astronomer was remarkable.

Hubble and Humason worked together for several years. Hubble set the scientific agenda, and Humason made the actual observations. Their collaboration started with a photographic survey of large numbers of galaxies, and then moved on to improving the measurements of redshifts and distances. Hubble's first breakthrough came in 1925, when he observed Cepheid stars in the spiral nebulae M33 and M31, and used the variable as standard candles to determine their distances by direct astronomical means.

Cepheids are a particularly useful kind of variable star for measuring cosmic distances. In 1908, Henrietta Swan Leavitt of Harvard College Observatory investigated thousands of Cepheids in the Large Magellanic Cloud. These Cepheids are all at essentially the same distance from us. She found a tight relation between the intrinsic luminosity of a Cepheid and its period of pulsation. Sixteen years later (1924) later, Hubble established the distance to Cepheids in the Andromeda galaxy, and showed that the variables were not members of the Milky Way. Hubble was able to estimate their true luminosities, that is to say, the absolute amount of energy they radiated as star light. This calibration, when compared to the *apparent* brightness registered on a photographic plate, gave the distance to the stars and thus to the galaxy containing the variable stars.

This was the commencement of calibrating the distance scale of the universe using a standard candle approach, whereby distances are determined by noting the apparent brightness of objects having known intrinsic brightness. Although Hubble's measurements were considerably in error by today's standards, they established beyond doubt that spiral nebulae were huge stellar systems far beyond our galaxy. With that job done, Hubble next resolved to use galaxies as tools for uncovering the large-scale structure of the universe.

Gradually Hubble saw the immensity of the universe unfolding before his eyes. In December 1924, at the 33rd Meeting of the American Astronomical Society in Washington, DC, Hubble announced a major finding concerning the spiral nebulae M31 and M33:

> Under good observing conditions . . . the outer regions of both spirals are resolved into dense swarms of images in no way differing from those of ordinary stars. A survey of the plates . . . has revealed many variables among the stars, a large proportion of which show the characteristic light-curve of the Cepheids.

Finding Cepheid stars in these systems and measuring their properties gave Hubble the standard candles needed to firmly establish how far away they were. The data showed that M31 and M33 were unquestionably extragalactic: he gave their distances as just under one million light-years, considered a stupendous distance in 1924. The island universe hypothesis had been confirmed.

■ A Cosmological Model to Fit the New Data: Enter, Georges Lemaître

Our account now moves geographically from Hubble in California back to Eddington's Cambridge, and from the fact finding at Mount Wilson to the meaning of those facts. At the commencement of the academic year 1923–24, the train station at Cambridge thronged with freshmen and returning undergraduates, almost all of them young men, hauling their trunks and suitcases. One figure stood out in this mêlée of privileged teenagers: Georges Lemaître, a handsome man aged twenty-nine with swept-back black hair, rimless eyeglasses, sporting the black clerical garb of a Catholic priest. He stayed at St Edmund's House, a residence for young Catholic priests studying at Cambridge. In this pleasant retreat, he passed the academic year in Cambridge, studying the application of general relativity to astronomy under the Quaker theorist, Eddington.

After attending a Jesuit school in Charleroi, Belgium, he had enrolled in 1911 at the school of engineering at the University of Louvain, to become a mining engineer. However, his interests and studies moved steadily toward pure mathematics. In August 1914, shortly after World War I began with the invasion of Belgium, he joined the Belgian army and served (in the front lines) throughout the war until its end. He was then discharged from military service on August 19, 1919, having been awarded the Military Cross with palms.

On resuming his studies at Louvain, he abandoned civil engineering for physics and mathematics, at which he excelled. Upon receiving his doctorate in 1920, he chose to study for the Catholic priesthood at a liberal seminary, where he adopted the reassuring philosophical concept that reason (science) and faith (theology) are two different roads to the truth. Importantly, for his future work in cosmology, Lemaître's natural theology encouraged him to consider that the universe may have had a finite beginning: it had not existed for all time.

Given Lemaître's rather ambitious goal to find a possible origin and evolution of the universe consistent with Einstein's theory of general relativity, it was natural for him to travel to Cambridge and study with Eddington as his mentor. The Cambridge year provided the start, and it took him another four years to achieve a breakthrough in his quest to use general relativity for modeling the universe. After the year with Eddington, he went to Harvard College Observatory, and from there to the Massachusetts Institute of Technology, returning to Belgium in 1925 (with a PhD degree) as a lecturer.

By the time he returned to Belgium, Lemaître had four key facts in hand for a new cosmology. First, from de Sitter he had the idea that the universe is expanding. Second, Hubble had shown him how the distances to nebulae could be found by observing their variable stars with the 100-inch telescope. Third, Slipher's spectra established that the spiral nebulae are extragalactic and gave their velocities. Finally, Lemaître's own doctoral thesis for MIT set out a suitable flexible model of the universe that could be adapted to fit the observations and still be consistent with general relativity. He was now in a position to use Einstein's toolkit.

What Lemaître sought were solutions to the field equations that would avoid the problems and oversimplifications of the two ex-

Figure 2.5. Georges Lemaître and Albert Einstein, on the occasion of their meeting at Caltech in 1933. (Archives Lemaître, Louvain, Belgium)

tant theoretical models—Einstein's static universe, which was full of matter, closed, and unstable, and de Sitter's model universe, which was open, expanding, and empty. Although both of those models were solutions to Einstein's equations, they could not be the correct solutions for our own universe: the first was catastrophically unstable and the second contained no matter in it! To break the impasse, Lemaître considered a situation "where . . . the radius of the universe is allowed to vary in an arbitrary way." He likened the universe to a gas in which "molecules are the extragalactic nebulae."

In April 1925, Lemaître had presented his interpretation of the recently reported recessional velocities of the nebulae to the annual meeting of the American Physical Society in Washington. He explained that these velocities were the result of the expansion of space, and that the speeds should be approximately proportional to their distances from us. This was four years *before* the publication of essentially the same idea by Edwin Hubble, who also based his conclusions on Slipher's data. In June 1925, Lemaître visited Hubble at Caltech, and also met Einstein. Lemaître's curiosity then led him to Flagstaff, where he carefully inspected Slipher's spectral results (see figure 1.4).

In a paper published in 1927, Lemaître announced the results for an expanding universe. He explained the recession velocities of the extragalactic nebula as the "cosmical effect of the expansion of the universe." This brilliant analysis, combining the new solutions to Einstein's equations with the new observations, was the first written exposition of modern cosmology. The paper was published in French by the Scientific Society of Brussels. By modern standards of scientific communication, Lemaître blundered by publishing in a language that U.S. astronomers did not read. American astronomers were completely unaware that a Belgian priest had derived a relationship (later called Hubble's Law) linking the distance to a galaxy to the speed by which it is receding. To do so, the cleric had used distance measurements that Hubble had published in 1926, together with the velocities of forty-three extragalactic nebulae, mostly those originally obtained by Slipher, and then publicized by

Gustav Strömberg, who was working at Mount Wilson. Lemaître found the rate of expansion of the universe to be only slightly different from the value published by Hubble at a later date.

In October 1927, Einstein was in Brussels, attending the Fifth Solvay Congress. This meeting of twenty-nine eminent physicists, seventeen of whom were or would become Nobel laureates, was on electrons and photons and devoted to the newly formulated quantum theory. Lemaître did not participate in this gathering of physics big wigs (he was not invited), but he took a train from Louvain to Brussels to meet Einstein, who was the youngest participant at the Congress. The two men, one with a high-winged collar and necktie (Einstein's famed rumpled look was the affectation of success and as a young man he dressed the part of a dandy) and the other wearing a Roman collar, strolled together along the tree-lined paths and round the lake of the Parc Léopold (see figure 2.5). Einstein commented favorably on technical aspects of the mathematics in Lemaître's paper, but then shocked Lemaître by remarking that from the point of view of physics, the notion of an expanding universe was an abomination that he could not accept.

Einstein's remark unsettled Lemaître, who nevertheless persisted with the idea that the expansion of the universe could explain the velocities that Slipher had observed among the nearby galaxies. After all, Lemaître had taken the trouble of meeting Slipher and looking at the original spectra in Flagstaff. Furthermore, Eddington continued to champion his former pupil's discoveries. It is not clear if Einstein's initial opposition to the idea of an expanding universe was based on some deep physical reasoning or, more likely, on unfamiliarity with the facts discovered at the western U.S. observatories.

Briefly stated, Lemaître the theoretician had seized Einstein's toolkit with both hands and brain, from which he constructed the first modern cosmological model. He fitted it to the preliminary observations that he gleaned from the published literature. It remained for Hubble, the consummate observer, to accumulate far better data and establish the cosmic law that would forever be associated with his name. He had not been idle.

■ Physical Cosmology and the Expanding Universe

Having established that galaxies lie far beyond the Milky Way, Hubble moved on to found the discipline subsequently known as physical cosmology. In California, thanks to the 100-inch telescope, he realized that observations of galaxies could actually map out the large-scale structure of the universe if he could estimate the size of a galaxy from its appearance. Then he could estimate its distance from us by its apparent (angular) size. He began this quest by looking at the various shapes of galaxies—ellipticals, normal spirals, barred spirals, and irregulars—classifying them in the manner of a nineteenth-century plant taxonomist. In this task he faced no equal, and certainly no competition. Hubble wanted to use the appearance (or "type") of a galaxy to estimate its distance. Here is an analogy to help us understand Hubble's method. If we see something the apparent size of our thumb at the length of our arm that looks like a giraffe, we know it must be much farther away than the object of the same apparent (angular) size that looks like a mouse.

The Hubble galaxy classification scheme (published in 1926) was an important first step in estimating the distances to the nebulae, provided one could tell from the appearance of a galaxy if it was a giraffe (large) or a mouse (small). The Hubble scheme does a fairly good job of enabling us to sort galaxies by size and is still used as a convenient descriptive tool. It sorted galaxies into different types according to shape: ellipticals, normal spirals, barred spirals, and irregulars.

He ordered the elliptical galaxies according to the degree of ellipticity, and used letters of the alphabet to arrange the spirals. About ten years later he presented the Hubble sequence (as it later became known) as the tuning-fork diagram. Elliptical and lenticular galaxies acquired the name "early-type" galaxies, while the spirals and irregulars were "late-type" galaxies. This confusing nomenclature established an erroneous belief that Hubble was speculatively proposing an evolutionary sequence. As a matter of fact, in 1927 Hubble stated:

> The nomenclature, it is emphasized, refers to position in the sequence, and temporal connotations are made at one's peril. The entire classification is purely empirical and without prejudice to theories of evolution.

During the mid-1920s Hubble next asked: is the large-scale distribution of the galaxies uniform? He tackled this question using the technique pioneered by William Herschel, who had been dead for more than a century. Hubble made counts of the numbers of galaxies brighter than a given apparent. More and more galaxies were detected, as he looked deeper and deeper to fainter and fainter objects. Looking out to larger distances will always produce more faint objects, because distant volumes of space are larger, the objects within them are more numerous but also made more faint by the greater distance. The number of objects seen to a given distance is proportional to the volume of space surveyed enclosed, and therefore the count scales with the cube of the distance. On the other hand, the apparent brightness of the galaxies must fall off as the square of the distance. Hubble found that the number of galaxies of a given magnitude increased with apparent magnitude

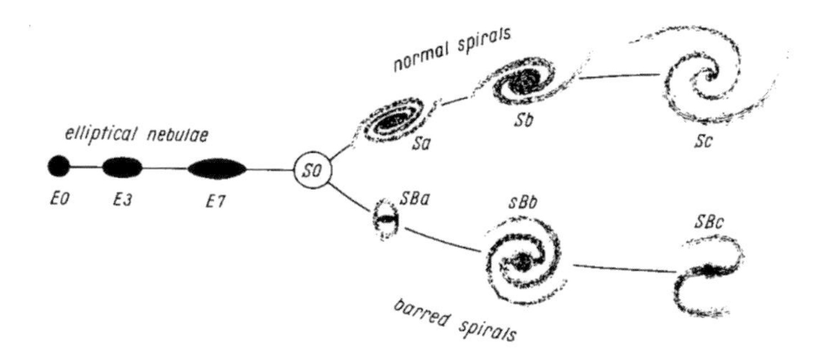

Figure 2.6. Hubble's scheme for the classification of galaxies. Hubble always regarded this as a taxonomic classification scheme. The notion that Hubble described it as an evolutionary sequence is an error made by later followers of Hubble's work. Most galaxies, including our Milky Way, are now known to combine elements shown separately in the Hubble scheme: a flat, spiral disk containing young stars and a fairly spherical ensemble of older stars. (From *Hubble: The Realm of the Nebulae*, 1936)

in exactly the manner expected for a uniform distribution. This test is brilliantly simple; it does not depend on knowing what the objects are or how bright they are. It only depends on knowing that the observer is not in a special place and that the mix of objects of different types and intrinsic brightness is everywhere the same, that is to say, that the universe is (statistically) uniform and homogeneous. In which case, the argument can legitimately be reversed. If counts of the number of galaxies and their magnitudes follow the expected relation in all directions, then in all likelihood we are looking out into a distribution of objects that is uniform and homogeneous in space.

The elementary exercise of counting objects seen on photographic plates, and noting how fast the numbers add up as one looks to fainter and fainter galaxies, was a breakthrough. William Herschel had started it, but he used stars rather than galaxies, and he was defeated by interstellar dust, which obscured distant stars. Although a relatively simple discovery, it had profound implications for cosmology, because it could henceforth be assumed that the universe is homogeneous and isotropic (the same in all directions).

What Hubble did next also had great cosmological significance. He worked out estimates for the typical masses of normal galaxies, from which he made a first estimate for the average density of the universe, the product of the number of objects per unit volume times the mass of each one. He realized that this parameter would have great significance for cosmology, because it would give a measure of the amount of gravitating matter that could slow the global expansion. We will see in later chapters how these estimates of the cosmic mass density grew over time.

Over the next three years Hubble assembled data to estimate the distances to twenty-four galaxies for which velocities had already been ascertained. To measure the distance to the nearest seven galaxies he used their Cepheid stars as standard candles. For the next thirteen, he used the brightness of the most luminous stars, and he used the brightness of gaseous nebulae for the four most distant galaxies. From this sparse sample of galaxies, boosted

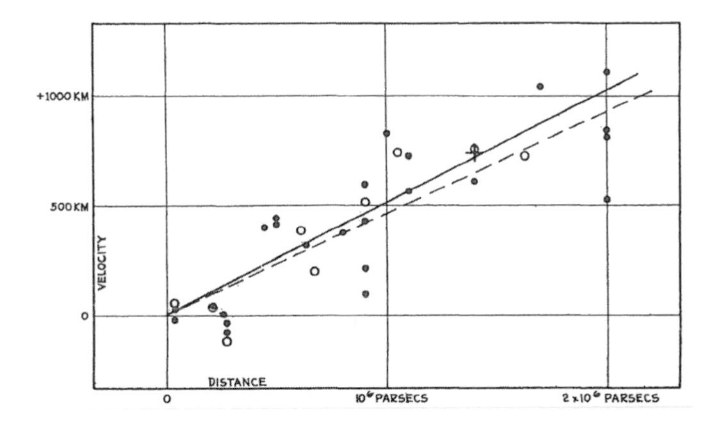

Figure 2.7. The Hubble diagram as first presented in 1927. More distant galaxies are receding from us at a greater velocity than those closer to us, a result that had been presented by Lemaître in the same year, the first published evidence for the expanding universe. These data are for 22 nebulae. Data points shown as black disks refer to individual nebulae. Open circles are the result of grouping and averaging nebulae in clusters. (From Contributions from the Mount Wilson Observatory, 1927, vol 3)

by data obtained by Slipher and the Humason, Hubble found the (approximately) linear relationship between velocity and distance that eventually provided the definitive evidence in favor of an expanding universe. Ironically, Hubble, as we noted, was not the first to write down what later came to be known as the "Hubble Law." Indeed, the Hubble law had been prefigured in Lemaître's neglected paper of 1927.

Hubble took a big risk in his 1929 paper, as the quality and quantity of the data only marginally supported the revolutionary conclusion of global expansion of the universe. Only a visionary would have thought of plotting a straight line through the scatter of points and deducing a *Law*, or even, as Hubble more modestly put it, "the empirical redshift distance relation." In his 1929 paper Hubble wrote:

> The results establish a roughly linear relation between velocities and distances among nebulae for which velocities have

> been previously published, and the relation appears to dominate the distribution of velocities.

In his next sentence Hubble disclosed that Humason had already commenced a program of observation at Mount Wilson to extend the survey to greater distances.

When Einstein was confronted with Hubble's results, which were in fact a good deal better than those that had been available to Lemaître two years earlier, he accepted the expansion of the universe, conceding that there was now no need for his proposed cosmological constant, which had been designed to allow a static (non-expanding) universe.

Hubble's result, that the recession velocity of a galaxy increased with its distance from us, in an apparently linear fashion, was actually somewhat troubling. Hubble realized that this trend could not persist forever. In an "Einstein universe" nothing can go faster than the speed of light, so a distance must be reached, the "radius of the universe" where the Hubble law must fail. He went on to calculate such a radius, and to estimate the total number of galaxies that would be within it. The specific cosmological model that he (inappropriately) adopted was the de Sitter model. But his calculation correctly estimated the distance over which relativistic effects would become important and redshifts would become large. The final paragraph of Hubble's fifty-two-page paper in the 1926 *Astrophysical Journal* is remarkable:

> The distance to which the 100-inch reflector should detect the normal nebula was found to be of the order of 4.4×10^7 parsecs, or about 1/600 the radius of curvature. Unusually bright nebulae, such as M31, could be photographed at several times this distance, and with reasonable increases in the speed of plates and the size of telescopes it may become possible to observe an appreciable fraction of the Einstein universe.

Putting it another way, Hubble concluded by saying that the 100-inch can photograph normal galaxies out to about 15 million light-years, and that that distance is just 0.15 percent of the radius of the observable universe. Bettering this required improvements in photographic emulsions and larger telescopes. Thus, Hubble's work with the 100-inch anticipated even greater discoveries that the 200-inch Palomar telescope and the aptly named Hubble Space Telescope would achieve.

The year 1930 marks what the historian Helge Kragh has identified as a paradigm shift in cosmology. Partly, this was due to the action Lemaître finally took early in that year in sending to Eddington an English translation of the 1927 paper. Eddington reacted favorably to the reminder from his former student. He sent a letter to *Nature* (published on June 7, 1930) praising the brilliant work of Lemaître, and he also penned a missive to de Sitter, telling the Dutch astronomer of "Lemaître's brilliant solution." De Sitter communicated the exciting news to Harlow Shapley. Thus did the professionals discover the pioneering physics work of Lemaître. The Belgian priest was becoming a cosmic celebrity. The theoreticians realized that Lemaître's theoretical model was consistent with both physics (general relativity) and astronomy (Hubble's data).

It is interesting to note that, contrary to some philosophical writings on the subject, there was no need to wait for the "believers" in the older views to be gradually convinced or to die out. Einstein was convinced rapidly, and it took only two years for the world of science to see how nicely the new picture cohered both internally and with the real world as observed by the new telescopes.

Lemaître's Synthesis Model Foretells the Contribution of Dark Energy

Furthermore, Lemaître's model solved another outstanding problem. Early in 1931, Eddington had given a popular lecture using the dramatic title "The End of the World: from the Standpoint of Mathematical Physics." This dealt with the so-called heat death of the universe, a concept from nineteenth-century thermodynamics in which a universe that lasted forever would ultimately run out of energy and everything would cool down to the same very low temperature, as its entropy approached a maximum value. Simply put, in a static universe, in which time stretches out forever in the past (and in the future as well), the stars would run out of fuel, whatever their fuel was, and ultimately they would blink out. We would find ourselves in a tepid universe peering out into a bath of the starlight from the expired stars. This, obviously, has not happened! Lemaître realized that this constituted an additional fatal flaw for the static Einstein model, and, when he came across the text of Eddington's lectures, he pointed out that a universe with a real beginning would not have this problem. In an expanding universe with a definite origin in time, the stars might not yet be old enough to have run out of their store of energy. There was no inconsistency in imagining a universe with a finite lifetime still having enough fuel left to keep the lights on.

In a letter to *Nature* (May 9, 1931), Lemaître wrote from Louvain, "the present state of quantum theory suggests a beginning of the world very different from the present order of Nature." He speculated that the universe might have commenced as a single quantum in which the expressions time and space had no meaning. While at this distance it is not possible for us to know what exactly Lemaître meant, it became known as the theory that the world started as "a primeval atom" and it stimulated the popular imagination of the time (even today one can download innumerable musical versions of "Lemaitre's Primeval Atom"). As we look back from the modern era, it strikes us that this picture has an uncanny resemblance to the inflationary birth of the universe (the hypothesized domain in which quantum and gravitational effects were competing) that we will discuss in our final chapter.

It was a heady era for wild cosmological inquiry, and the champagne flowed (but not in the United States, where prohibition ruled). Eddington arranged for Lemaître to speak in a session at the British Association for the Advancement of Science in October 1931. The giants of cosmology also made contributions: an elderly

Figure 2.8. Lemaître in the 1930s. (Copyright Bettman/CORBIS)

de Sitter, Eddington, Sir James Jeans, and Arthur Milne, the professor of mathematics at Oxford.

The event was a stunning success, attracting a crowd of about two thousand. Lemaître introduced a "fireworks theory" of evolution into cosmology. He proposed that the primeval superatom, whose existence he had postulated, had disintegrated, producing cosmic radiation and ordinary matter in a spectacular fireball. It is more than a little unsettling to note how close all of these speculations are to the best current physical theory, tested and calibrated by satellite observations of the cosmic background radiation field that was not to be discovered until decades later.

To transform his additional ideas from rank speculation to solid science, Lemaître sought a more detailed mathematical formulation. By assigning a positive value to the cosmological constant (or, putting it another way, by introducing a source of cosmic repulsion), Lemaître found a model that expanded smoothly from an initial state at the beginning of time. Far from the cosmological constant being Einstein's blunder, Lemaître firmly believed in it. He showed remarkable foresight at the U.S. National Academy of Sciences meeting in November 1933, when he interpreted the operation of the cosmological constant thus: "Everything happens as though the energy *in vacuo* would be different from zero."

Even the vacuum of the hypothetically empty universe contained energy, the pressure of which could lead to the expansion of space-time. While we are no further along today in understanding the cause of the vacuum energy, the best evidence available as of the writing of this book strongly supports Lemaître's revival of Einstein's cosmological constant and its interpretation as a manifestation of a hidden energy, latent in the universe. It was George Gamow who later claimed (in his 1970 autobiography) that Einstein had called the introduction of the cosmological constant into the field equations the biggest blunder of his life. In the 1931 edition of his popular book on relativity, Einstein added an appendix explaining why this fudge factor was no longer required. But, as we noted, in a series of strange twists in the progress of science, the cosmological constant, invented by Einstein for the wrong reasons, and developed by Lemaître, and then again discarded by the community of cosmologists, would return dramatically at the end of the twentieth century as the primary component of the present universe under the sobriquet "dark energy."

By now Lemaître really was a celebrity, and his visit to the United States in 1932–33 aroused great interest. The fact that he donned rather severe clerical dress added to the mystique; here was a Catholic priest who had found the origin of the universe. His 1933 meeting with Einstein, who was by then a convert to this new, globally expanding model of the universe, received sympathetic press coverage. James E. Peebles, probably the leading cosmologist of the postwar epoch, has written of Lemaître: "According to the usual criterion for establishing credit for scientific discoveries Lemaître deserves to be called the Father of the Big Bang Cosmology."

■ Hubble's Achievements

Edwin Hubble was a more comfortable celebrity. When Einstein visited Caltech in August 1931, Edwin the showman took him up the twisty road to Mount Wilson in a sleek Pierce-Arrow touring car. The expanding universe had been treated in the popular press as a problem for Einstein's theories. But after his visit to Mount Wilson, Einstein welcomed the expanding universe. Hubble had shown him new results that swelled the database for the velocity-distance relationship.

But, as we noted, in a series of strange twists in the progress of science, the cosmological constant, invented by Einstein for the wrong reasons, and developed by Lemaître, and then again discarded by the community of cosmologists, would return dramatically at the end of the twentieth century as the primary component of the present universe under the sobriquet "dark energy." Although Hubble later became feted as the father and discoverer of the Big Bang, the truth is, as we have seen, that Vesto Slipher had discovered the redshifts of galaxies, and Georges Lemaître had known of the correlation between redshift and distance.

Hubble's truly monumental achievement was the establishment of the extragalactic distance scale; and in this he was the scientific entrepreneur par excellence. His results on M31 and M33, followed by the Hubble Law, rocketed him to A-list celebrity status in Hollywood. He and his wife dined with Douglas Fairbanks, they showcased Mount Wilson to Cole Porter, and, according to British cosmologist Fred Hoyle, entertained visitors with English afternoon tea in the Rose Garden of the Huntington Library.

Hubble, in the dozen years from 1924 to 1936, established the foundations of modern observational cosmology, solving four of the central problems, any one of which would have guaranteed him a position of the first rank in the history of science. In our opinion, only Galileo outshines Hubble as a supreme example of observational achievement in cosmology. Hubble's four notable achievements are these:

- Hubble proposed the standard classification system for nebulae, both galactic (diffuse) and extragalactic. The galaxy classification system has become the Hubble morphological sequence of galaxy types.
- With his discovery of Cepheids in M33 and M31, Hubble settled decisively the question of the nature of the galaxies, whose correct solution, to be fair, had previously been given using what many believed to be inconclusive arguments.
- From 1926 to 1936 the distribution of galaxies, averaged over many solid angles, was determined to be homogeneous in distance. This proved that galaxies were a major component in the structure of the universe, and it demonstrated that observations of galaxies had the potential to reveal cosmological parameters. Galaxy counts to the magnitude limit of the Mount Wilson 100-inch telescope were then used to attempt a measurement of the radius of curvature of space.
- The linear velocity-distance relation was set out in a discovery paper in 1929, followed by a series of papers with Milton Humason between 1931 and 1936 that verified and extended the relation to higher redshifts. This discovery led to the notion of the expanding universe that is central to the cosmological models of the present day, together with an age for the universe then estimated at 2 billion years.

Hubble hardly ever dabbled with theory. He was happy to leave analysis and speculation to the theorists. He felt that the photographic and spectroscopic techniques of his time were not sufficiently sensitive to allow observations that could distinguish different cosmological models. In his 1936 popular account, *The Realm of the Nebulae*, Hubble's final paragraph reads as a mission statement of all astronomical observers:

> The search will continue. Not until the empirical resources are exhausted, need we pass on to the dreamy realms of speculation.

What were the speculations? Cosmology as a subject of investigation had not really been born yet. Certain high priests of speculative general relativity had made mathematical models, but the astronomical world was barely paying attention; an examination of the literature in the 1930–50 period finds almost no interest in using astronomical data to determine the right world model. Most must have agreed with Hubble that the data were not good enough to warrant such an exercise.

But one simple point was clear. If the expanding universe meant that the cosmos was becoming less and less dense at the present time, then the data did strongly indicate that the universe must have been much more dense in the past. If one were to imagine the clock running backwards, then there was a moment, some finite time ago, a time in the past comparable to the age of the Earth, when the galaxies would all have been crashing together. And if at that time the galaxies had not yet been made, then the matter from which they were to be made would have been imploding (as we look at it in the rear view mirror). Or, running time forward, the matter must have taken part in some giant explosion. The Lemaître model might be right or it might be wrong, but the simple seventh grade lesson that distance equals velocity multiplied by time for moving objects required there to be some explosive event in the past. This was an unnerving thought. It seemed almost biblical and unphysical, but it was the clear implication of the accumulating astronomical data. More data were needed.

■ Big Science to Attack the Big Problem

Spectacular discoveries at Mount Wilson had shown that extragalactic research needed mountain sites for telescopes and big money to finance them. During the first half of the twentieth century, European astronomers entirely lacked the ambition to construct large telescopes on mountain sites. Even as late as 1948, the British, under the leadership of their Astronomer Royal, showed a marked reluctance to board passenger ships in order to spend weeks on end travelling overseas. Of course, the astronomers in California used their automobiles to get to the telescope: the 100-inch was a one-hour drive from Pasadena, Lick Observatory was perhaps two hours by car from San Francisco.

In 1928, George Ellery Hale secured a grant of $6 million from the Rockefeller Foundation, for "the construction of an observatory, including a 200-inch reflecting telescope ... and all other expenses incurred in making the observatory ready for use." In the period 1930–34 Hale searched locations for the planned 200-inch telescope. Sites in Arizona, Texas, Hawaii, and South America were considered, but the early favorite and the eventual winner was a site at an elevation of 5,600 feet on Palomar Mountain, 100 miles southeast of Pasadena, California. For the telescope mirror, Hale first tried fused quartz. When that failed, he approached the Corning Glass Works in 1932 with the challenge of fabricating a 200-inch mirror using the low thermal expansion borosilicate glass (its trademark name is Pyrex). The first blank cast failed (it had voids), and it now resides in the Corning Museum of Glass. The second blank took one year to cool down, and was delivered in 1935. The next year construction on site commenced, only to come to a sudden halt during World War II. The 200-inch saw first light in 1948.

In contrast to the previous decade, essentially nothing of much importance happened in cosmology in the dozen years from 1936 to 1948. However, there was one major exception during World War II, when a German-born astronomer at Mount Wilson, Walter Baade, made a very great discovery concerning stellar evolution, and subsequently revised (that is to say corrected) Hubble's value for the age of the universe.

Baade had arrived in the United States from Hamburg, Germany in 1929 on a Rockefeller scholarship that enabled him to visit Mount Wilson. Two years later, the observatory offered him the staff position that would change his life. When Hitler declared war on the United States on December 11, 1941, Baade became an enemy alien, who was confined to Los Angeles County. His colleagues at the observatory were drafted to weapons development programs, so Baade faced little competition in using the 100-inch. In April 1942, Los Angeles introduced a military curfew, corralling Baade in his house from dusk to dawn.

Walter Adams, the director of Mount Wilson, appealed against this restriction to the highest possible level in Washington, with the result that the army command issued an exemption allowing Baade out at night on the strict understanding that this was only for professional purposes on Mount Wilson. He therefore had almost sole access to the 100-inch, and the wartime brown-out of Los Angeles gave him optimal observing conditions that would never again be experienced at Mount Wilson.

Under very dark skies during the fall of 1943, Baade used newly available, very sensitive photographic plates to examine M31 (the Andromeda galaxy) and resolve individual, red giant stars. This required immense skill as an observer. In the central region of M31, Baade found that the brightest stars were yellow supergiants, whereas in the galaxy's spiral arms the giants were red and blue. Baade had discovered that the spiral arms contain young stars, which he named population I, the brightest members of which were the red and blue giants. But the central regions of spiral galaxies and globular clusters were different, and he named them population II. The immediate puzzle was to understand what physical mechanism could sift and sieve stars into two very different populations.

Baade shared this news of the two populations of stars with an excited (and excitable) Fred Hoyle, who visited Pasadena in late 1944. Later still, Hoyle and others would explain that the population II stars were among the first generation of stars formed. It was Baade, too, who on learning of Hoyle's interest in novae gave him a hot tip: "If you are interested in such things then why not look at supernovae, which are vastly more powerful." Hoyle did exactly that when he got back to research at Cambridge in 1946 and worked on the origin of the chemical elements. Baade's discovery of the stellar populations, and the interest he sparked in supernovae, were later to have profound influence on observational cosmology.

■ The Steady State Model Universe and the Big Bang

The year 1948 saw a shake-up in cosmological thinking because of the publication of the steady state theory by the Cambridge trio: Hermann Bondi, Thomas Gold, and Fred Hoyle. In the developing standard picture the universe expands and, although the details were not clear, it must expand from an explosive event before which time and space had no meaning. Hoyle thought this was absurd and in violation of every physical law that was known. The steady state theory was the antidote that he proposed. This theory, in its time a rival to the Lemaître picture, postulated that the universe does not evolve, but on the largest scales looks more or less the same, no matter what the location or epoch of the observer. How does one avoid in this picture the thinning out of the matter density as the universe expands and the associated heat death? A new law of physics was postulated; the expansion of the universe would be compensated by the continuous creation of matter to fill the expanding voids.

Figure 2.9. Fred Hoyle, Herman Bondi, and Thomas (Tommy) Gold at the Eleventh General Assembly of the International Astronomical Union, Berkeley 1961. (Carvel Gold)

The proponents of the steady state theory argued that this steady creation of matter from nothing, at all places and times, was no worse than the implicit creation of everything from nothing in a "Big Bang." Hoyle coined that picturesque term in a radio broadcast on March 28, 1949. He explicitly denied choosing the expression as an insult. Although the steady state theory never achieved much traction in the United States (except among journalists), in the UK, Hoyle received so much publicity in the popular media in the 1950s and 1960s that observers felt compelled to confirm the big bang theory with solid results. And by now they had the 200-inch telescope with which to voyage deeper into space. The hope was that it could see far enough out in space and far enough back in time to discriminate between the steady state model and the (recently labeled) big bang model.

When the 200-inch came on stream in 1948, Baade seized his opportunity. His first program was a detailed study of Cepheid variable stars in M31. He took several hundred photographs of the highest quality on which he identified 500 Cepheids. He passed the identifications to his collaborator Henrietta Swope, who obtained the periods and light curves. When Baade failed to resolve RR Lyrae stars in M31 (they are fainter stars than the Cepheids, but they should nevertheless have been visible with the 200-inch telescope) he became suspicious of Hubble's published estimate for the distance to the galaxy. If this distance should turn out to be larger than Hubble's estimate, then the whole chain of reasoning on which the cosmic distance scale was based would be blown off course: all other distances would have to be adjusted upward as well. Those looking at the cosmological implications were also becoming suspicious. They were understandably troubled by the discrepancy between Hubble's computed age for the universe (the elapsed time since the moment in the past when the galaxies would all have overlapped), only 2 billion years, and the longer ages calculated for the oldest rocks on Earth. The latter was estimated, with little chance of error, from dating methods based on the known rate of radioactive decay of unstable chemical elements. If all the extragalactic distances were larger than had been thought, then there was more time available in the past.

This is a recurring problem in astronomy: the universe must not be younger than the objects that live within it. During his eventful career Baade published few papers, preferring instead to converse with the incessant stream of visitors passing through Pasadena, and also by engaging in scientific meetings. In September 1952, the International Astronomical Union (IAU) convened in Rome for its triennial jamboree. Baade, by then in his sixtieth year, had saved up two treats for his Roman audience.

The IAU has always taken care to record faithfully the deliberations of its members. In Rome, Baade chaired a formal session of presentations on galaxies, and asked Fred Hoyle to take the minutes that would later be published in the official volume of proceedings. As Hoyle later recalled, Baade pointed out a simple error that Hubble had made in relating the brightness of the two kinds of variable stars: Cepheids and RR Lyrae. So, the data that Hubble had used to gauge the distances to galaxies now stood corrected. The Hubble age of 2 billion years was too small. Baade announced an age of 3.6 billion years, breaking at a stroke a logjam that had for fifteen years frustrated the development of explosive models of the universe. This estimate is at least comparable to the age of the Earth but still not as large as we now know it to be. Current methods of several different types have pretty much nailed the age of our universe to be 13.75 billion years, with an error estimated to be only 0.8 percent.

Baade's second surprise concerned the new science of radio astronomy. During World War II, big metal dishes had been used to send out radio waves that would bounce off approaching warplanes, the returning signal being recorded by the same dishes. Then after the war it was realized that these same radar dishes could be used as telescopes to record radio waves arising from outer space, and thus the new discipline of radio astronomy was born. By 1951, the first list of cosmic radio sources had been made by radio telescopes at the University of Cambridge. Of course, it was not clear whether these radio sources were nearby stars, or something else beyond the Milky Way. At the 1952 IAU meeting, Baade fished from his pocket a small glass slide with the optical spectrum of a radio source in the constellation Cygnus. The spectrum had been taken at Palomar by his Caltech colleague Rudolph Minkowski, using the 200-inch.

The slide showed a recession velocity that implied a distance of hundreds of millions of light-years, thus demonstrating at a stroke that the radio sources were probably all located in galaxies at cosmological distances, rather than associated with unusual stars in the Milky Way. Martin Ryle, head of radio astronomy at Cambridge and a pioneer in mapping the sky in radio waves, later said of Baade's showmanship, "from that moment we knew that we were in the cosmology game." That was because some of the radio sources he and his colleagues had discovered were visibly associated with distant galaxies emitting enormous amounts of energy as radio waves.

One year after the Rome IAU, Edwin Hubble died, passing the baton to Allan Sandage, who became responsible for developing the observational cosmology program for the Mount Wilson and Palomar telescopes. Under Sandage, the goals were recalibration of Hubble's extragalactic distance scale and finding good standard candles—objects of fixed brightness.

Sandage used the 200-inch to great effect in order to measure the rate of expansion the so-called Hubble constant H_0 (the ratio between the velocity of a distant object and its distance from us), and how that constant is changing with time. It is obvious that the matter in the universe, acting gravitationally, will cause the expansion rate to slow (the detailed physics is given in the next chapter), so the universe must be decelerating. Sandage's measurements allowed him to compute the "deceleration parameter," q_0, a measure of how rapidly the expansion is slowing. These were the two numbers required by the then standard model of cosmology. Finding them out from observational studies would, astronomers believed, completely specify both the past and future of the observable universe. Of course we now know that this picture was far too simple. And both the additional problems and their solutions were to come from a quite different direction. At the same time, it had dawned on radio astronomers in Europe and Australia that their new way of exploring the cosmos, through radio waves rather than light, could contribute to cosmology. We develop these themes in chapter 4.

The half-century of observational discovery since 1900 had embedded our still valid picture of the Sun and the solar system living in a flattened galaxy of billions of stars into a much grander universe. Kepler had dethroned us from being at the center of the solar system to the role of living on a small satellite planet revolving with others around our Sun. Then Herschel and others made it clear that our magnificent Sun was a rather ordinary star, although its central place in the universe was still in the textbooks at 1900. Observations by Shapley moved it many thousands of light-years to a position circling the galactic center along with the bulk of the stars in our spiral galaxy. And then Curtis, Hubble, and others showed that our galaxy was one of many in an enormous expanding cloud.

What an extraordinary transformation of our vision of the world. The explanation of how this happened and what was to happen next would use the cosmological models that were developed from Einstein's general relativity. The solutions to Einstein's difficult equations, which were found by Lemaître and Friedman, were now known to be appropriate for the expanding universe in which we seemed to live. Big scientific programs at the major new observatories in the western parts of the United States were attempting to determine which of the new solutions provided the best fit to the real world. In our next chapter we show how the essential physics of that expanding universe can be followed with high school physics, and then how the physics and the observations were combined into the unified picture that we call the modern paradigm of cosmology: a flat, hot, big bang model dominated by dark matter and dark energy.

Chapter Three

Let's Do Cosmology!

■ The Big Bang: A Starting Point That Cannot Be Escaped

The fact that we live in an expanding universe had been amply confirmed. The implications for the future were obvious. But, what of the past? Were the arcane speculations of the Belgian priest, Lemaître, to be taken seriously? Is the true but intimidating general theory of relativity the only key to understanding the evolution of the universe and to extrapolating from the present back into the past? In fact, the clear conclusion that there was an explosive event in the past is easily shown, and much of the basis for modern cosmology can be easily presented. A simplified version will be presented in this chapter, with the details reserved for appendix 1.

When the authors of this book were young, there were popular accounts by Eddington, Jeans, and others that not only told the story of the astronomical discoveries of the era, but also showed, by direct logic and simple mathematics, how these results could be derived. We would like to show here how the latest results of cosmic science can be understood and not just accepted on faith by anyone who enjoyed high school math. This small chapter, and the associated appendix, is our own attempt to provide a simple, quantitative discussion for the evolution of the cosmos, much as those earlier authors did for stars; so some simple physics equations will be used. Perhaps most important of all, we will outline how certain metrics were constructed that allowed cosmologists to predict by inference, from observationally determined numbers, whether in the future our universe will expand or collapse.

But, the reader may ask, does one not need Riemannian geometry and tensors to address the issues of cosmology with any precision? As we have noted, Einstein was not able to formulate general relativity until he had learned that new and difficult branch of mathematics. Almost a decade passed until Alexander Friedman and Georges Lemaître found the first, full solutions of Einstein's equations in the 1920s. Those solutions are still the bedrock of mainstream cosmological thinking, and the derivations of these classic solutions are formidable, but we will see that we do not need full solutions of Einstein's equations to predict the future of the universe.

To study the universe as a whole it is obvious that the full, general theory of relativity is needed. For the comprehensive picture, we need to consider regions that cannot be seen directly, regions so far away that they are beyond the horizon from which light can reach us. Thus, it was long thought that to solve the equations of cosmology, one needed the complete relativistic theory—the entire, rather difficult, Einstein toolkit. But let us see if we can reproduce much of the Friedman-Lemaître solutions ourselves, using only elementary physics and mathematics. Specifically, can we formulate the question that we really want answered: What will happen to our local piece of the universe—does it expand forever or does it re-collapse?

This leads us to ask a simpler set of questions. Instead of trying to solve for the structure of the whole universe, let us look at something more manageable, one very small piece of the universe—say our own galactic neighborhood. Is this an easier problem to formulate and solve? Yes it is. If one formally derives the relevant equations needed to understand the expansion or contraction of only one small bit, one does not find in them the quantity c, the speed of light. Newton's laws are enough. From the absence of c it

Figure 3.1. The Sombrero galaxy (M104) is 50,000 light-years across and 28 million light-years from Earth. It was thought to be a luminous disk of gas surrounding a young star, until 1912, the year in which Vesto Slipher measured a velocity of 700 miles per second. Seen nearly edge on, the Sombrero is classified as a spiral because of the dusty disk that is seen, but if it were viewed face on, it would probably be considered a giant elliptical galaxy. (NASA and the Hubble Heritage Team STScI/AURA)

is clear that relativity is not really needed for local forays into the universe. The intimidating apparatus of Riemannian calculus is not needed. We can go back to high school mathematics to study the one small representative piece and thus to find out how the whole universe behaves.

A little bit of thought shows that one in fact *should* be able to do the exercise with no more than Newton's physics and high school mathematics, although, as far as we know, the elementary treatment of cosmology presented in this book has not previously been published. The vital point to remember is that, in a globally homogeneous universe, every little mass element behaves exactly like every other one at a given instant in cosmic history. Therefore, a careful enough study of the fate of one piece can tell us a great deal about the evolution of the whole.

But before embarking on this journey, let us first see why everyone, even the detractors of the big bang theory, realized that Hubble's discoveries implied that there must have been an explosive event in the past. Consider any two galaxies in Hubble's universe. We live in one of them, the Milky Way, and the other, say the famous Sombrero, is distant from us by distance D. It has velocity V with respect to us. How long a time, T, has it been traveling? This is an exercise that we all had in seventh grade, and the answer is well known: the distance traveled is the velocity multiplied by the time: $D = V * T$. But, now we note that Hubble found that the velocity a galaxy is receding from us is proportional to its distance from us, and we call the constant of proportionality Hubble's constant H: $V = H * D$. These two simple equations can be put together to give $D = H * D * T$; and then, dividing both sides of this result through by D gives the simple result: $T = 1/H = 15$ billion years. *Voila!* If we want to know when the Sombrero galaxy was on top of us, we just take the reciprocal of the Hubble constant. That is pretty simple. We measure one number and we have the answer. But there is something puzzling here. That one number has nothing to do with the Sombrero galaxy. And if we did it for any other galaxy in the universe, we would have obtained the same answer. All of the galaxies were on top of us 15 billion years ago.

It is true that we have neglected gravity and all kinds of other complications in this simple exercise, but it is worth pausing to think about the answer. Suppose that instead of standing on the Milky Way to start, we were living in the Sombrero galaxy and made the calculation starting there. Well, obviously, the answer would have been the same. In fact, wherever we started, we would have come to the same conclusion. The whole universe of galaxies would have been on our heads about 15 billion years ago. The straightforward implication of the Hubble law is inescapable. Something happened roughly 15 billion years ago that can only be described as a vast cosmic explosion.

What have we omitted in this too simple treatment of cosmology, and how does that affect the answer? Most important, we have neglected gravity. As we explain in appendix 1, a stone thrown upward against the force of gravity slows steadily and typically reverses course and falls again to Earth. It is true that if we throw something upward with enough speed, while it will slow down somewhat, it will escape the Earth. Similarly, if the universe started out in a giant explosion, then the gravity due to the matter in the universe would certainly have slowed the rate of expansion somewhat. Was the gravity enough to make the universe stop expanding and start to collapse again? Apparently not, since it is still expanding at a rather smart rate. Is the gravity enough to make it slow down in the future, stop its expansion, start to contract, and then to collapse in a horrible implosion to a black hole? We do not know the answer, but we do know how to formulate the problem and what measurements we should make to see what to expect.

■ Observational Cosmology, the Biggest Puzzle to Be Solved with the Biggest Telescope

Allan Sandage set out to solve this problem observationally with the 200-inch Palomar telescope. Essentially he looked out a great distance to earlier times to see how rapidly the universe was expanding in past epochs. Hubble and then Baade had determined the Hubble constant at the current epoch, now. With a bigger telescope, could he not determine its value at great distances from us in the past? We saw that, measured locally, the current age of the universe equalled $T = 1/H_0$. If that was true in the past as well, then at earlier times, when the age, T, was smaller, H would have been larger. If the universe was decelerating, then clearly the Hubble "constant" would be still larger in the past. How much larger was it then? In any case, sufficiently careful measurements on a distant piece of the universe should reveal how Hubble's constant had changed since earlier times. By a quantitative measurement of the deceleration, Sandage aimed to determine the "deceleration parameter" q_0 (please see appendix 1, equation A15) that would tell us the answer. If this arcane parameter were found to have a value less than ½ at the present time, then we could rest assured that the universe would expand forever. If it were found to be greater than ½, then a "big crunch" would await us in the future. Sandage found,

Figure 3.2. Allan Sandage, protégé of Hubble and founder of modern observational cosmology shown at the Palomar Observatory 200-inch telescope, completed in 1949. (Photo copyright estate of Francis Bello/Photo Researchers Inc.)

by doing the observations as carefully as he could, that q_0 was between 0.5 and 1.0 with a fair amount of uncertainty. From this he concluded that we apparently live in a "closed" universe that is doomed to collapse in the future.

But there was another, more direct way to approach the question of the future. One could simply measure the mass of a typical galaxy, and multiply by the number of galaxies that were found per unit volume, which was something that Hubble had measured. That calculation yields the mass density of the universe. A high value of that number would mean that gravity, due to the observed masses in the universe, would ultimately cause the cosmos to recollapse, and a small value of the average mass density would mean that gravity was relatively unimportant and that the expansion would continue forever, perhaps slightly slowing but never stopping. The parameter that was invented to represent the average mass density in the universe was based on a Greek letter, the capital omega, and was written as Ω_{matter} (please see appendix 1, equations A6 and A8). If the parameter Ω_{matter} was found to be small compared to one, the universe would expand forever, and, if it was bigger than unity, then the universe would collapse. Sandage's results on the deceleration parameter implied that when Ω_{matter} was determined, it should be found to be significantly bigger than one.

We show in the appendix that, if our understanding of cosmology is correct, then the two cosmological parameters that we have defined in this chapter are equivalent: knowing one enables us to compute the other by the simple equation (again, this is in appendix 1, equation A17): $q_o = ½\ \Omega_{\text{matter}}$. Thus, in the cosmological models that Hubble and Sandage were exploring, there were only two independent parameters that needed to be measured: the Hubble constant H_o, and the average mass density Ω_{matter}. If they could be determined observationally, then (given the mathematics presented in the appendix) the whole past and future of a representative piece of the universe would be determined. In figure 5.7 we show how a small spherical piece of the universe evolves with time, either growing steadily or expanding to a maximum size and then collapsing, and figure 7.1 shows a diagram indicating how larger or smaller values of Ω_{matter} give radically different future trajectories. Of course, all of the models have a big bang, an explosion, in the past. The exact elapsed time, T, since that event does depend on the specific model, and it is always a bit less than our first naïve estimate of $T = 1/H$. While it was realized that this could be problematic, the bigger issue was to find a consistent overall picture for the future evolution: eternal expansion or re-collapse.

From the 1960s onward, many observational astronomers set out to measure Ω_{matter}. They were shocked to find that the initial results showed it to be always far below unity: the galaxies simply did not weigh enough to cause much deceleration. Were we missing something? Was there some other component, mass not in galaxies, that could alter the calculus? It was more than a little troubling that the two ways of doing the exercise were giving diametrically different answers.

Sandage's efforts to directly measure the deceleration pointed to a closed universe that would likely re-collapse; but direct measurements of the mass density by the Dutch astronomer, Jan Oort, and others indicated that there was far too little mass in galaxies for this to happen. This situation persisted up to and through the 1960s. Much debate and head scratching ensued. Was there some error in either the calculations or the measurements? Astronomers made inventories, adding up all the planets and stars, including all of the gas and dust that they could find, and they determined that this was only a few percent of the magic number: from direct observations Ω_{matter} had a value of about 0.03. Cosmologists concluded that the universe was open, implying that it would expand forever. The simple metaphor of throwing a ball up in the air (or shooting up a rocket) and trying to guess if it will ultimately fall down again or if it has enough velocity to escape the Earth is a pretty good model, and we return to it in chapter 7. By our best measurements the velocity of expansion was ample to keep the universe expanding forever.

Half a century ago, the suspicion was always present that we were finding a lower bound to the truth with regard to the cosmological, average density of matter. There could always be more matter in some hard to detect form that was being overlooked. Thus, although the conclusion was reached from studying the mass density in all observable forms that Ω_{matter} was far less than unity, this conclusion was held to be suspect, since the inventory of matter might be missing important components. Newspapers talked of the "missing matter" as if there were a priori knowledge that really Ω_{matter} must be unity. The components that people wondered about were stars too faint to be seen or gas between the stars that had a temperature and density that made it hard to observe. They were not thinking about the dark matter that was soon to be discovered and found to be much more abundant than all the stars and gas put together.

When the great enterprise was initiated at the world's preeminent observatories, Hubble and Sandage knew that it would not be easy to find exactly how much bigger the Hubble constant was in the past than at the present. Their research would tell them how effective gravity had been in slowing the expansion, and this could be used to forecast the future behavior of the universe. To say that this was an ambitious undertaking would be to wildly understate the difficulties. At the time the exercise began, the local value of the Hubble constant was poorly known and was uncertain by more than a factor of two. Given that, how would observers be able to say that in some barely detected, faint, and distant part of the universe this parameter was a few percent bigger than it was measured to be locally and that, furthermore, the small change in the Hubble constant detected over the eons of time was securely measured? The audacity was impressive. The largest and most costly telescopes at the world's best sites were to be used to make these measurements and to ascertain the fate of the universe!

The quest was an exciting story for the media. The general public followed the to and fro as pundits opined on one side or the other of the issue. Significant funding was expended on the effort, which stretched over continents and decades. Of course it failed. It failed for several different reasons. First, the measurements were simply too difficult to make with the required precision. Second, the observers did not take into account that the objects they were looking at in the far away universe would not be exactly the same as similar ones seen nearby. We return to the effects of cosmic evolution in chapter 5. And third, they had the wrong model for the universe, because they were not including in their calculations the effects of Einstein's cosmological constant.

■ The Grand Project Was Initially Too Difficult

Thus it was believed that the evolutionary course of the universe could be determined by two equivalent but observationally independent methods. Astronomers could add up the mass in a little volume, measure the rate of expansion of that volume, and see if the gravity due to this much mass is sufficient to ultimately stop the expansion and cause a re-collapse. Or they could make careful measurements of the Hubble constant and its rate of change to determine the answer by purely geometrical methods. The two methods should, of course, give the same answer. Unfortunately, there were major impediments to this grand project, which was the focus of the astronomical community for roughly half a century, each one of which turned out to be nearly fatal. We will now examine each of them in turn.

1. Problems with the Hubble constant. The quantities that enter into a direct measurement of the Hubble constant or the rate of change of the Hubble constant or the critical density are notoriously difficult to measure. When the grand project was begun, it was thought that standard candles and standard meter sticks could be found in astronomy. In reality the search for reliable yardsticks turned out to be phenomenally difficult. The cause of the problem is now easy to see, but back then it was overlooked. Whatever objects we use as our standards (such as bright galaxies) clearly were not existent at the beginning of time. So they must be part of an evolving population. Therefore they are not standard––unchanging—quantities. But that evolution is just what we are trying to measure. We are in a trap of circular logic. Many clever techniques were developed to overcome this problem of circular reasoning, but it has always come back to trip us up. At the present time much faith is being vested in the constancy of the properties of certain supernovae (a special kind of exploding star), and they really are good standard candles.

But the history of cosmology teaches us there is little doubt that more careful investigation could show that these too evolve with cosmic time. If so, they cannot be used for precision measurements unless we can separately understand that evolution. Galaxies—the building blocks of the Hubble and Sandage approaches to cosmology—*must* evolve, if only because they are made out of stars, and those stars will change in appearance as they burn up their fuel. This now obvious point was made in the early 1970s, by a young woman astronomer, Beatrice Tinsley, who had a large role in bringing down the house of cards with her insistence on the reality of galactic evolution. But we have more to say on that later.

2. *Problems with the matter density.* The measurement problems of the matter density, which gives an independent way of determining the fate of the universe, are not any easier. We can only measure what we can see, and what we can see is always an incomplete picture of what there is. Consequently, our measurements of the density of matter are always too low. We can see and can count the bright stars, but what of the faint ones? What of the fainter binary companions of massive stars, of the low mass stars, of the planets, comets, dust, and gas in the universe? The list goes on and on. Efforts to assemble an inventory of the whole list of astronomical objects have been heroic, and they have produced moderately persuasive results. As we shall see later, the bright stars represent perhaps only a tenth of all mass in stars. The gas that never was turned into stars is yet another factor of ten more; but even being liberal with all estimates, it was found that the total was far below the magic, critical mass density, ρ_{crit} that would give $\Omega_{matter} = 1$ (appendix 1, equation A8). And what of the mass that cannot be seen? Is there something else? Alas, yes. It was later found (although the "pre-discoveries" were in 1937 and they were ignored) that the universe is pervaded by "dark matter," and that this stuff dominates over ordinary matter by roughly a factor of ten! We return to this story in chapter 6.

3. *Is the local universe special?* What if the little volume that we are studying is not a representative piece of the universe? What if it is significantly underdense or overdense compared to the average. In the former case, it is part of a void, and astronomers see these everywhere in extragalactic space. In that case we will measure too rapid an expansion, or, we will find too little matter in the sample volume compared to the average over the universe. We do see significant fluctuations in the density of galaxies all around us—clusters and voids, super-clusters and super voids. Where does it all end, and how big a volume must we survey before we can be sure that the studied sample is truly "representa-

tive"? These are questions without simple answers. From the study of the cosmic background radiation we have an idea of what is happening on the largest observable scales, and many of the new ground and space-based optical, radio, and X-ray surveys are beginning to approach the depth required, but these questions of how big is big enough are going away only slowly over the decades of observation.

4. *And what are we to make of dark energy?* The whole enterprise that we have embarked on in this chapter is predicated on the belief that Newton's laws tell us all that we need to know, that the force of gravity is the only one that we need to care about. That too has turned out to be wrong. Although we do not understand it at all, it seems as if there is another force, which is so puny that it is immeasurably small on the scale of the laboratory or even the solar system. This force makes space act as if there were little springs in it laid end to end—everywhere! This force dominates even over gravity on the very largest scales. Its name is "dark energy." But what's in a name? Giving something a name, even if it makes us feel more comfortable, tells us nothing. And we have not a clue as to what the dark energy is. We return to the tortured discovery of this strange substance in chapter 7. But what we will learn there is that our equation of motion needs an extra term in it to allow for this bizarre extra force in the universe.

We have succeeded in our mathematical excursion and it carried us to the level of understanding reached about two-thirds of the way through the last century. Using simplified (but accurate) versions of the tools from Einstein's toolkit, we derived not one but two different quantities that observational astronomers could pick up and use to measure global properties of the universe in our vicinity. One was essentially a balance sheet. However, the inventory was not of financial assets but of the total mass density of the universe in terms of its galaxies, stars, planets, gas, dust, and so on. The second tool used careful measurements of Hubble's constant nearby and at earlier epochs to show us how rapidly gravity was causing the expanding universe to decelerate, to slow down.

If our equations and measurements are accurate, either tool could tell us the future fate of the universe. And we have a check: they should both provide the same answer. One method, used by Allan Sandage, seemed to indicate that re-collapse was to be expected, but within the immense uncertainties of the measurements the results might allow the open expanding model, which was strongly preferred by the other method. But, more seriously they gave quantitatively quite different, really discrepant answers (by roughly a factor of thirty), thus providing a clue that some force other than gravity was acting on cosmic scales. By the strictest standards of science, the then standard cosmological model had been tested and found to be false!

However, that is taking us too far ahead in our cosmic saga. What we have done in this chapter and the appendix is to show that with relatively simple math only, and with just Newton's laws, one can write down the conditions for the future of the universe in a way that should allow astronomical measurements to accurately predict that future. Einstein's equations and Riemannian calculus are not needed. There is something grand and wonderful about this, that we, poor creatures that we are, dwelling on an inconspicuous piece of dust called Earth in a quiet corner of the universe, can (or at least believe that we can) make calculations and measurements from our local experience that we believe should accurately predict the fate of the whole. It is quite an impressive exercise in what the Greeks called hubris, is it not?

Chapter Four

Discovering the Big Bang

■ Did Our Universe Have an Explosive Birth?

In astrophysics the connections between the very small and the very large are intimate. Atoms are very small and are composed of a much smaller and extremely dense nuclear ball made up of neutrons and protons surrounded by a cloud of electrons. This chapter tells how the knowledge that physicists obtained of the workings of atomic nuclei transformed our understanding of both stars and the cosmos. In all of the familiar chemical reactions made in our laboratories on Earth the atomic nuclei are untouched, and the atoms interact and combine by sharing electrons in the clouds surrounding the nuclei. The nuclei and hence the chemical elements themselves are unchanged during any chemical interaction. But in extreme circumstances, when nuclei collide with great force, these normally quite stable, central balls of particles can combine or fragment with the release of enormous amounts of energy. This can happen on a small scale on Earth in specially designed accelerators, in a grand scale in the centers of stars and, as we shall see, on the cosmic scale in the early universe. Prodded by work related to World War II, this branch of physics made gigantic progress in the 1940s. The astrophysical spin-offs from that work explained to us not only how and why the stars can shine, but how the universe itself must have started in a vast cosmic explosion.

When Hubble found that the universe of galaxies was currently in a state of expansion, that led us immediately to the question broached in the last chapter: if things are flying apart now, then was there not a time, rather recently in cosmic units, when it was very, very compact? As we showed in the previous chapter, the simple extrapolation backwards in time indicates that there must have been some singular event roughly 10 to 20 billion years ago. This is not an immense interval compared to the 3.7 billion-year age of the Earth. So the question was forced upon us: how did the universe begin? It is as if we were looking at the debris flying outward from a giant explosion in the past, a "fireworks universe" was how Lemaître put it, or "big bang" as Fred Hoyle termed it in 1949. Could that simple-minded explanation be correct? And if so, at early times, was it hot or cold, smooth or lumpy?

The questions are endless, and we could make up any number of fascinating and entertaining scenarios for the past evolution of the universe, which just about every culture has done, but how would we know if any particular picture was right, scientifically correct? Could cosmology, the comfortable, speculative myth-making activity we have engaged in for ages, become a real science? As we noted in the preface, to be a proper science it is necessary that the cosmological scenarios we come up with are made sufficiently definite, clear, and mathematically precise, so that they can be tested empirically. The tests should be precise enough and clean enough so that a theoretical picture can be proven to be incorrect. The theory must be "falsifiable"; it must be possible to observationally or experimentally show that it is wrong.

As it turned out, two technologies—nuclear physics and radar—which were developed for quite other purposes during World War II, held the keys. Understanding nuclear physics gave us the methods to compute the cooking of the chemical elements that would have occurred consequent to the giant explosion that came to be called the "*hot* big bang." Radio astronomy gave us the tools to detect and measure the radiation left over from that explosion, first from the ground and later from space.

Einstein's famous equation $E = mc^2$ held the key to understanding the fueling of the Sun's enormous energy output, and makes it clear that similar nuclear processes must have happened when the temperature of the universe was as hot (or hotter), in the distant past, than it is today in the interior of the Sun. This provided an explanation for the origin of the common, light chemical elements, abundant in stars and the interstellar and medium. The cosmic cooking (or transformations of the chemical elements) occurred in an incredibly dense and hot radiation field that could not have totally dissipated. The argument was that the enormous amounts of radiant heat would have been diluted by the cosmic expansion but should still be visible if careful searches were made.

Although investigations were started to look for the remnant "cosmic background radiation field," it actually was discovered in a totally accidental manner. And then the pieces all came together. Hubble had written what we now realize was the second act of our cosmic drama. The dramatic first act was now envisioned in all its fiery splendor. The primordial global explosion, the formation of the chemical elements, and the stage settings for the later world of galaxies had been discovered.

Historically, the origin of many of the chemical elements in the big bang explosion was not the first problem in astrophysics to be addressed with the new nuclear physics. There was a more urgent question, one that had been troubling scientists since it was formulated by Lord Kelvin in the late nineteenth century.

■ What Makes the Stars Shine?

Nuclear astrophysics had its modern genesis in the 1930s with the nuclear physicist George Gamow, the flamboyant immigrant from the Soviet Union, who became the leader of the new game. He was born in 1904 at Odessa, then in the Russian empire. He attended universities at Odessa and Petrograd (named Leningrad in 1924, and regaining it original name St. Petersburg from 1991), where he studied under Alexander Friedman, the brilliant mathematician who had discovered and classified the cosmological solutions to Einstein's equations. From 1928 to 1931, Gamow worked on nuclear physics and stellar physics in Copenhagen and Cambridge before returning to the Soviet Union, where he found life disagreeable. So, when he and his wife traveled to Brussels in 1933, for the Sixth Solvay Conference in Brussels, which was devoted to nuclear physics, they promptly defected. In 1934, he was appointed a professor at George Washington University, where he worked with Edward Teller, who would later be known in the press as the father of the hydrogen bomb.

Figure 4.1. Niels Bohr Institute Copenhagen, 1930. Those on the front row (*left to right*): O. Klein, N. Bohr, W. Heisenberg, W. Pauli, G. Gamow, L. Landau, H. Kramers included some of the founders of quantum mechanics and nucleosynthesis. (Niels Bohr Archive, courtesy AIP Emilio Segre Visual Archives, Margrethe Bohr Collection)

Gamow and Teller organized several meetings in Washington, DC, in the late 1930s which focused on understanding the fueling of the stars. The energetic requirements were so enormous that they knew it would require the transformation of elements by nuclear processes and would utilize Einstein's classic relation between mass and energy. The first tentative steps in applying nuclear physics to astrophysics were taken at a meeting in the late 1930s, where the central topic had been nuclear reactions in the interiors of stars, Gamow's preoccupation at the time.

Teller once referred to the topic as "Gamow's game," although it was the nuclear scientist Hans Bethe who became the game's key player, and who would receive the Nobel Prize in 1967 as the result of his efforts. Bethe solved the central problem of how stars, including especially our own Sun, can continue to shine for billions of years, steadily emitting copious amounts of luminous energy, when any known source of chemical energy, such as our fossil fuels, would be used up in a tiny fraction of the stellar lifetime. It was known that normal helium contained four nuclear particles, two protons and two neutrons, so in principle it might be formed from collisions among the four protons of four normal hydrogen atoms. It was understood by this time that the two principal chemical components of the Sun were hydrogen and helium, so it seemed possible that the required nuclear reactions could be occurring in the Sun. Also, one helium atom weighs less than four hydrogen atoms, so combining the latter to make the former would liberate an enormous amount of energy. Figure 4.2 shows schematically the "proton-proton cycle" discovered by Bethe that converts hydrogen to helium. Variants of this process are being investigated today to utilize thermonuclear fusion as a source of energy for electric power at many labs on Earth.

The idea was not entirely new. Almost two centuries prior to this, an English chemist, William Prout, became the first modern scientist to address it. In 1816, he proposed that all elements had formed from hydrogen by a process of coagulation. A century later that hypothesis intrigued Eddington, who suggested in his popular book *Stars and Atoms* that helium nuclei were made by com-

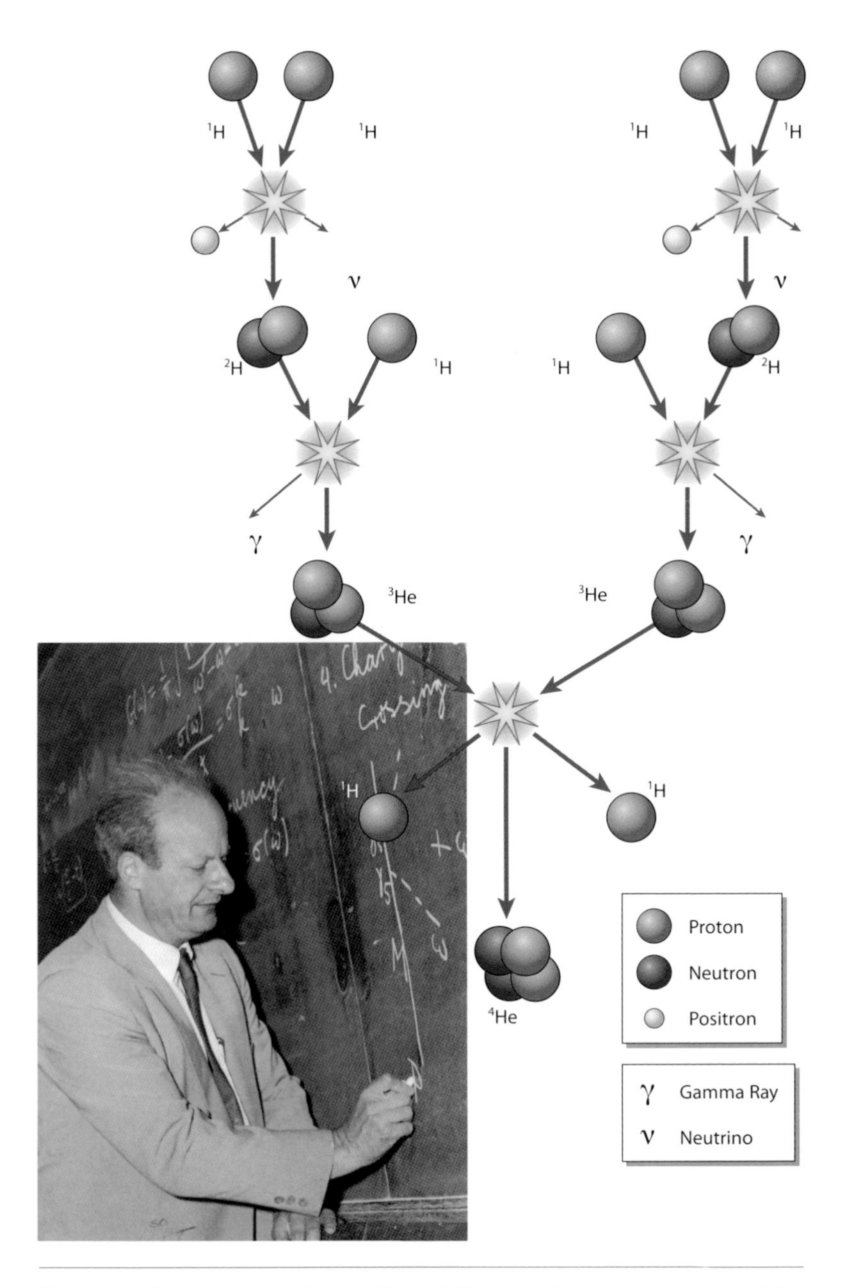

Figure 4.2. Hans Bethe at the blackboard. Diagram shows how 4 protons (the nucleus of hydrogen atoms) combine to make one helium nucleus. (Edith Michaelis, courtesy AIP Emilio Serge Visual Archives)

bining four hydrogen nuclei during an unknown process that involved loss of mass. But how was this to be done? What was needed was a detailed pathway of nuclear reactions that would work under the conditions of astellar interior. No one had found the route.

Hans Bethe had fled Germany in 1933. After a stint in the UK at the University of Bristol, he settled on the rugged and isolated campus of Cornell University, where he quickly established himself as the world's foremost authority in nuclear theory. He knew nothing about the interiors of stars when he attended the 1938 Washington meeting organized by Gamow and Teller, but rapidly became interested in the generation of energy by the Sun, which he approached as a problem in nuclear physics. Bethe, together with Charles Critchfield, a former student of Gamow, and Teller, proposed in 1938 the stellar energy scheme now known as the proton-proton cycle whereby hydrogen atoms, the main chemical constituent of the Sun, could sequentially combine to make helium via a several-stage process, potentially releasing energy at a high enough rate to account for the enormous solar luminosity. Traveling by train back to Cornell, Bethe worked out the basics of this cycle, which followed exactly what Eddington had intuited: the fusion of four hydrogen nuclei to make one helium nucleus. Six months later, and now thoroughly knowledgeable in astrophysics, Bethe devised the alternate CNO cycle, a more complex catalytic process to reach the same end. In both cycles the net output after several stages is one nucleus of helium-4 and two neutrinos. The energy released through these fusion processes is so enormous that it gives stars like the Sun a lifetime of 10–15 billion years, of which less than 4 billion years have expired. Bethe had answered the child's questions: "what makes the Sun shine?" and "will it ever turn off?"

Bethe, then, was the founder of a new branch of astrophysics, in which the origin of stellar energy was turned into laboratory science, a practice that after World War II was extended also to cosmology, which took nearly half a century more to develop into a precision science. He became the pioneer at the interface between particle physics and cosmology, and the American nuclear physicists dominated cosmology for some time. During the war years he was the head of the theoretical division of the Los Alamos Laboratory and made major contributions to the development of both the A-bomb and the H-bomb, subsequently campaigning effectively against bomb tests in the atmosphere and for the treaties that moderated the nuclear arms race. Freeman Dyson called Bethe the "supreme problem solver of the twentieth century." The excitement of astrophysics at this time was clearly attracting the world's leading scientific minds.

■ Nuclear Astrophysics Moves to the Cosmos

Today, the seemingly independent disciplines of cosmology and nuclear physics are interdependent. The creation of that link began in March 1942 (four months after the attack on Pearl Harbor), when two dozen nuclear physicists and astrophysicists gathered for three days at George Washington University, four blocks from the White House, to discuss "The Problems of Stellar Evolution and Cosmology." Gamow, as the joint convenor of this eighth annual Washington meeting on theoretical physics, had suggested the topic.

The meetings in the late 1930s had been successful in their efforts to understand the fueling of ordinary stars but led to the conclusion that the origin of the chemical elements was to be found elsewhere. Bethe, for example, had this to say: "It is quite possible that the formation of elements took place before the origin of stars, in a state of the universe significantly different from today's." We now know this to be true for the lightest elements. In 1926, Eddington had written: "We do not argue with the critic who urges that the stars are not hot enough for this process; we tell him to go and find a hotter place." Nuclear physicists were also following a clue discovered by Cecilia Payne and Henry Norris Russell at Harvard, who had shown that the proportions of the different chemical elements were rather uniformly distributed among the stars, which suggested a common origin for the elements. Attention thus switched from stellar interiors to the early universe.

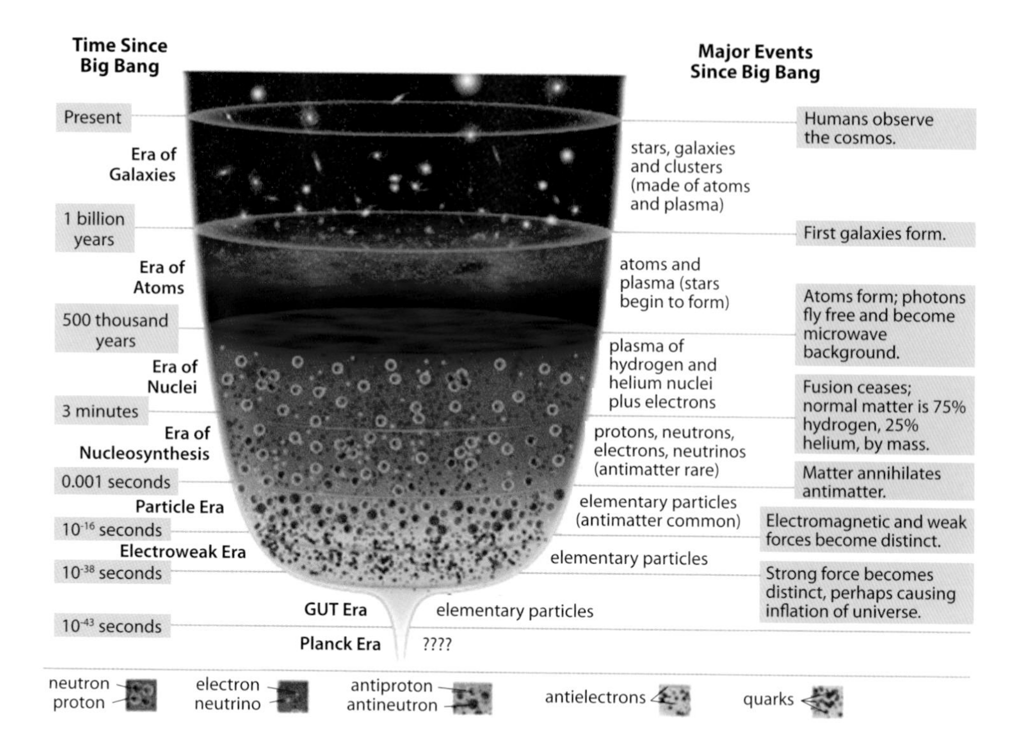

When the nuclear *astro*physicists, as they now styled themselves, convened at the 1942 Washington meeting, perky Gamow took the lead on the second day, April 24, which was devoted to the expanding universe and the origin of the chemical elements. The participants first noted that there continued to be a serious obstacle to the acceptance of the expanding model of the universe—that Edwin Hubble's value for the age of the universe was only one-half the geological age of the Earth. Gamow skillfully dismissed this marked discrepancy by questioning some of Hubble's assumptions, a brave thing to do given the latter's godlike standing. In his official report Gamow wrote:

> It is, however, desirable to retain the hypothesis of the expanding universe, since it provides a basis by which a great many phenomena may be explained. The most important of these is the riddle of the origin of the chemical elements—a

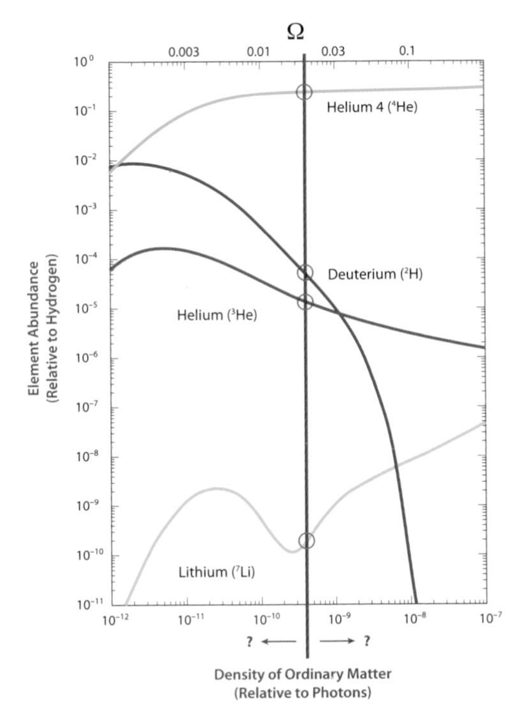

Figure 4.3. A modern view of the hot big bang model proposed by Gamow (*facing page*). The creation of the light chemical elements during the "nuclear cooking" of the red section of this figure fits well with the detailed, measured abundances of those elements (*this page*) and gives strong confirmation of big bang model.

process requiring high temperatures and densities such as could have existed only in the early stages of the expanding universe.

In his presentation he boldly stated that the elements had originated in an explosion that marked "the beginning of time" and resulted in the present expansion of the universe. Many of the participants at the meeting shared this point of view, even though the enthusiasm for the existence of an explosive epoch was the result of sheer imagination, influenced perhaps, as we shall see, by other currents in the real world. Gamow now fully endorsed the big bang picture together with the suggestion that the chemical elements had been created in a compressed primeval state.

Three years would elapse before Gamow took the ideas further (there was a war in progress) and he perhaps underestimated the importance of what he had achieved. In October 1945, Gamow wrote a letter to his friend Niels Bohr in Copenhagen, congratulating him on his sixtieth birthday. Gamow revealed in the letter that he had resumed work on the origin of the elements:

> I am . . . studying the problem of the origin of the elements at the early stages of the expanding universe. It means bringing together the relativistic formulae for expansion and the rates of thermonuclear and fission reactions. . . . the period of time during which the original fission took place must have been less than one millisecond.

This extract shows that in late 1945, Gamow had the essential idea of a *hot* big bang model that combined the Friedman-Lemaître equations (which described the physics of the expanding universe) with nuclear physics. On October 13, 1946, Gamow sent a short paper to *Physical Review* in which he combined the two strands of thought. He speculated that the early universe had consisted of a gaseous soup of neutrons. In essence, the entire early universe was a very large atomic nucleus made only of neutrons, rather like Lemaître's model of 1931, although Gamow gave the Belgian neither credit nor citation. Essentially, the two belonged to different scientific clubs. The crux of Gamow's proposal was that heavy elements were built up by the accumulation of neutrons, followed by beta decays (spontaneous nuclear reactions during which a nucleus emits an electron) that converted neutrons to protons. Gamow's paper also contained the exaggerated claim that elements could not be synthesized in stars. Outside of Gamow's group, however, many physicists in Europe felt that the stars held the key to the origin of the bulk of the elements.

Following the suggestion of Walter Baade to concentrate on very hot stars (red giants), Fred Hoyle in Cambridge obtained reasonable agreement between his calculations and the observations of the abundances of heavy elements. He looked at nucleosynthesis in the heart of red giant stars, a late stage of stellar evolution, that the Sun will reach in several billion years, where very high central temperatures pertain. So the question of the origin of the chemical elements—whether they were synthesized in the big bang of the early universe as a whole or in the smaller bangs of exploding stars—was still wide open.

Gamow's 1946 paper in *Physical Review* is sometimes hailed as the foundation of modern big bang cosmology, despite the fact that his startling conclusions received little support. Gamow clung to the idea that the essential process in building most of the elements found in nature must be the capture of neutrons by protons and nuclei in the early universe. The next move forward was made by Ralph Alpher, a doctoral student supervised by Gamow and working on primordial nucleosynthesis.

At a meeting of the American Physical Society, Alpher listened intently to a ten-minute presentation on the properties of neutrons. The data, from the Argonne National Laboratory, had been obtained to assist in the design of nuclear reactors. Alpher thought he could use them, not for reactor design, but in the design of the universe. He requested a full set of these data from Argonne, which supplied them immediately. Crucially, Alpher noticed immediately that the data on high-energy particle-particle collisions correlated with the relative abundances of the elements. In other words, the collisions that, according to nuclear physics, were most likely to occur, produced the chemical isotopes (the same chemical elements with different numbers of neutrons in the nuclei) that were most abundant on the Earth and in the Sun. This seemed to indicate that Gamow's basic idea was correct. The conclusion about the physical processes occurring was correct, but it did not provide evidence as to whether this process was happening in massive stars or had happened in the early universe.

The concept of explosive nucleosynthesis at early times now occupied Alpher and Gamow to the exclusion of everything else. Some of their deep conversations took place at Little Vienna, a bar on Pennsylvania Avenue where hard liquor apparently softened the task of understanding the universe.

As the months passed, Alpher became increasingly confident that he could model the creation of helium from hydrogen in the first five minutes after the sudden onset of the expanding universe (see the red area of fig. 4.3, p. 110). The ratio of hydrogen to helium so produced matched reality (see fig. 4.3, p.111), close to the observed ratio of hydrogen to helium in most stars. If Alpher was correct, most of the helium seen in the surface layers of the Sun had been made in the Big Bang, but in the solar interior there was a further transformation occurring via the processes that Bethe had discovered that would leave the Sun ultimately with an almost pure helium core. So, even though Alpher could not make progress at all with the heavier elements, by predicting the correct ratios for the formation of hydrogen and helium and their isotopes, he made a highly significant step, at a time when he was still a graduate student.

Gamow announced this breakthrough with a paper in *Physical Review* in 1948. He submitted the paper, titled "On the Origin of the Chemical Elements," but there was a catch. A legendary humorist, he could not resist the temptation to add to the paper the name of his friend Hans Bethe. Thus it was that the published authors of the paper were Alpher, Bethe, and Gamow, a famous pun on the letters of the Greek alphabet: alpha (α), beta (β), and gamma (γ).

Young Ralph, not surprisingly, took exception to the stealthy insertion of a big name who had contributed not one iota to the paper. As the student who had done all the computational hard slog, he resented being eclipsed by one of Gamow's japes. Bethe, for his part, was blissfully unaware of Alpher's feelings, and none of them had any idea that the alpha-beta-gamma paper would become one of the most important in the history of cosmology. But Gamow, ever playful and a heavy drinker, patched up the disagreement with a small celebration. He produced a bottle of Cointreau, poured generous glasses, and the three toasted the universe. He had doctored the label on the bottle to read not Cointreau but *ylem*, an obsolete noun meaning "The primordial substance from which the elements were formed."

■ The Fireball in Which the First Chemical Elements Were Made

The alpha-beta-gamma paper on the origin of the light elements became a milestone in the history of cosmology because calculations on nuclear processes had shown that the abundances of hydrogen and helium were entirely consistent with a fiery, explosive origin of the universe. Alpher and Gamow were the first to remark that hot matter ought to be accompanied by electromagnetic radiation called "thermal blackbody radiation." In an early triumph of quantum mechanics, Max Planck (in 1901) had correctly predicted the detailed spectrum (that is to say, distribution of colors) of the radiation that would be found within a cavity surrounded by walls at a definite temperature. The hotter the walls, the more the color shifts from red toward blue and ultraviolet.

Gamow estimated that in the early Big Bang, the nuclear cooker would be filled with very high energy radiation having at that epoch a mass density greater than the mass density of the nuclear soup itself. The density and temperature would be so high that, even after eons of expansion and cooling, the radiation might still be detectable. Fifteen years later this prescient suggestion would lead to a pivotal moment in the history of cosmology, when Arno Penzias and Bob Wilson detected the fossil radiation from the Big Bang, the primeval explosion within which the light elements had been cooked.

Ralph Alpher and Robert Herman took the next bold step, also in 1948. The nucleus of normal hydrogen is a single proton, but there is a common isotope, deuterium, having a nucleus that combines a proton and neutron in a tight bond. Gamow had already noted that the production of deuterium in the Big Bang could only have proceeded if the temperature in the fireball was about one billion degrees, 10^9 K. Alpher and Herman asked what the temperature of that radiation would be today: the thermal radiation necessary for element production would still be in the universe, but vastly cooler and diluted by the enormous expansion that had taken place over time. By making some plausible assumptions (and some lucky guesses) they calculated that the fossil radiation from the Big Bang would have cooled to about 5 K at the present time. This is strikingly close to the value measured many years later, 2.725 K.

Big bang nucleosynthesis, as originally described by Alpher and Gamow, lasted only seventeen minutes and took place throughout the universe. It began three minutes after the initial fireball, at which time the universe had cooled sufficiently to allow protons and neutrons to freeze out, and it ended about twenty minutes after its initiation because the density and temperature of the expanding universe fell below that required for nuclear fusion.

Critics of the fireball universe hit back, by pointing out the failure of the theory to explain element creation beyond the first two elements in the periodic table. Later development of the theory did show that traces of the elements lithium, beryllium, and perhaps boron were produced as well. The abundance of deuterium in the universe today is an important indicator of the conditions just after the Big Bang because deuterium cannot easily be made anywhere in the universe today: essentially all that we now see emerged in the furnace of the Big Bang. Importantly, in 1964, Fred Hoyle and Roger Tayler provided support for the theory by showing that the universal helium abundance could only be accounted for in terms of a ubiquitous synthesis everywhere in the universe during Gamow's big bang. It is a testament to Hoyle's scientific integrity that he publicized this important work despite the fact that it cast doubt on his own favorite cosmology, the steady state model.

The conclusive work on big bang nucleosynthesis was accomplished just after the discovery of the microwave background in 1965 by P.J.E. Peebles at Princeton, who wrote two papers on the expectations for the abundances of the helium isotopes (see fig 4.3, p. 111). These began a train of work that provided some of the strongest support for the big bang model, by showing that the detailed isotope ratios of several of the light elements—hydrogen, helium, and lithium—fit so well with the predictions of the nuclear cooking scenario.

Only the lightest elements could be made in the Big Bang. This led to a bottleneck: nucleosynthesis could advance no further in the fiery furnace of the first few minutes of time. The heavy elements—everything we find on Earth in abundance except the hydrogen in water—must have been made elsewhere, after the Big Bang. The solution of this puzzle would require a shift of the manufacturing site for the heavier chemical elements from the early universe to massive stars, a move that Hoyle had already made. In the late 1950s he, with two other British astrophysicists, Margaret

Figure 4.4. Fred Hoyle, with Geoff and Margaret Burbidge, relaxing at a picnic somewhere in the Rhone Valley, France, July 1958. (Photograph Thomas Gold, courtesy Carvel Gold)

and Geoffrey Burbidge, and the Caltech experimental nuclear physicist, William Fowler, conclusively demonstrated several of the detailed paths by which massive stars, late in their evolution, could make (and expel in giant "supernova" explosions) the bulk of the common chemical elements. While there are still numerous, detailed unsolved problems in this field on which astrophysicists continue to work (in particular, producing the explosions that accord well with observational data), the overall set of processes is well in hand.

The importance of the light elements for cosmology is that their formation process is easily understood, and their abundances carry information on the physical conditions during the first few minutes of the existence of the universe. Furthermore, the fact that their present-day measured abundances agree precisely with what is computed from the nuclear cooking during the Big Bang provides very strong support for the modern cosmological paradigm. The story of the early universe is brilliantly set out in Steven Weinberg's popular account, *The First Three Minutes*, published in 1977.

■ Direct Radio Observations of the Big Bang Fireball

The intellectual landscape in cosmology during the 1950s was as follows. Conceptually, there were several competing cosmological models in addition to the hot big bang. One model invoked oscillating universes, in which a cycle of expansion and contraction occurs. Another model was the aesthetically interesting, steady state cosmology proposed by Hoyle, Bondi, and Gold, back in 1948. In this scheme the expanding universe is infinitely old, and, as the galaxies move apart, new matter is created spontaneously, hence an alternative name for this cosmology: continuous creation. In the steady state cosmology the universe has the same appearance throughout time and space, and shows no signs of evolution. In the period 1950–65, a guerrilla force of steady state cosmologists, led by Hoyle, skirmished with an army of big bang enthusiasts. Needless to say, the correct prediction of the hot big bang model for all of the isotopes of hydrogen, helium, and lithium, which the big bang model could boast, was absent in the steady state model. This nontrivial defect of an otherwise appealing picture was amplified by evidence of a new and surprising sort that essentially confirmed the big bang model.

In the 1950s, the practitioners of the new science of radio astronomy joined in the cosmological fray when they discovered that their telescopes could detect radio emission from remote galaxies. Martin Ryle's group at Cambridge had made this breakthrough when the double radio source Cygnus A was found to be centered on a disturbed galaxy. The extragalactic radio sources, or radio galaxies, offered a means of discriminating between an unchanging steady state universe and an evolving big bang universe. The question being asked was this: does the distant universe have the same general appearance as the nearby universe? In all cosmologies, the distant universe is observed at an earlier epoch than is the nearby universe, because of the length of time—billions of years—for light and radio emission to reach our telescopes. The faintest objects in a survey tend to be those farthest from the observer. If the very distant regions had the same population of sources in them as the nearby universe, then in the steady state model, radio telescopes should find exactly the same relation between the number of objects found in the sky of a lower and lower brightness as found by Hubble at Palomar. In our previous chapter we mentioned that the number of sources counted should increase as one looked to fainter and fainter magnitudes, going as the –3/2 power of the observed brightness. This had been used by Hubble to demonstrate local cosmic uniformity and could be used on the much grander cosmic scale using the radio counts, since radio could be detected to much greater distances than optically observed galaxies.

Ryle's radio astronomy group, based in the physics department at Cambridge, decided to put steady state cosmology to the test. Despite being colleagues in the same university, Ryle and Hoyle had found it impossible to work together, the observer and the theorist striving to put the other down, sometimes even in public. Ryle attacked the problem fervently. He studied and counted the distant radio sources, concerning which the steady state had made a definite and falsifiable prediction: that the number of the objects per unit volume should be exactly the same at all times and places in the universe, and the counts should increase with decreasing brightness exactly as predicted by the theory. Ryle's attack involved counting the number of radio sources in different categories of brightness. This was Hubble's method, but applied to radio galaxies. Ryle conducted four distinct campaigns to survey radio galaxies.

Ryle's first two campaigns, in the early 1950s, seemed to disprove the steady state picture. At least, that's what Ryle felt. But the contrarian Hoyle maintained (and quite correctly, as it turned out) that, because of instrumental problems, the first two surveys were so flawed that they proved no such thing. The third and fourth surveys of radio sources were carried out at Cambridge with meticulous care and they did indeed show clear evidence for evolution in the universe. Hoyle by then was refusing to accept these results on the grounds that Ryle, originally an engineer, was incapable of cosmological observations. Ironically, from Hoyle's per-

Figure 4.5. Antony Hewish and Martin Ryle discuss a contour map of a radio source, 1974. (AIP Emilio Segre Visual Archives, Physics Today Collection)

spective, the 1974 Nobel Prize for Physics went to Martin Ryle and his colleague Tony Hewish (the discoverer of pulsars) for their contributions to the development of radio astronomy. By 1960, to the eyes of almost all astronomers, other than Hoyle and his coterie, the steady state theory looked totally implausible on two counts: the radio source numbers found by Ryle did not fit the steady state model, and the abundances of the lighter chemical elements did fit the big bang model. But the observers still lacked the killer punch that would give undisputed and conclusive proof of the hot big bang.

As in a good mystery story, the truth eventually emerges in an unexpected manner. From the mid-1950s onward, astronomers were certainly thinking about possible ways to detect fossil radiation from the Big Bang. Since the expected temperature of the radiation would be somewhere below 10 degrees kelvin, any observations would have to be made in the very short microwave segment of the electromagnetic spectrum, and it was obvious that any signal remaining would be extremely weak. An attempt to detect weak cosmic microwaves with an antenna would require high-gain amplifiers (to boost the detected signal) and low noise (so as to eliminate microwave radiation generated locally by the antenna and its electronic circuits). Although the British radio engineers, who invented radar for early detection of attacks by German aircraft, had first developed the technology during the war, the United States was now ahead in microwave technology.

In the United States, the ATT Bell Laboratories at Holmdel, New Jersey, had developed high-gain microwave antennas and low-noise amplifiers for research into radio communication. The availability of this high-quality equipment for radio astronomy would prove crucial in the eventual detection of the microwave background by Arno Penzias and Robert (Bob) Wilson. Note however that continuous development and improvement at Bell Labs was primarily driven by the need to demonstrate the feasibility of communication using satellites. NASA made important contributions to that development, but neither Bell Labs nor NASA intended to be players in the cosmology game at this stage.

Arno Penzias joined the radio astronomy group at Columbia University as a graduate student in 1958, after which he obtained a temporary appointment, in 1961 as a microwave radio astronomer working with engineers at Bell. His first task was to learn how to use a 20-foot microwave horn antenna to follow astronomical radio sources as the rotation of the Earth caused them to track across the sky. The practical experience gained would later be used to track communications satellites, but the work schedule left Penzias plenty of time to pursue radio astronomy with the same antenna.

Bob Wilson meanwhile had entered graduate school at Caltech in 1957. He learned how to build instrumentation for radio astronomy, and his doctoral thesis involved mapping the radio emission from the plane of the Milky Way galaxy. Wilson attended Fred Hoyle's cosmology lectures at Caltech, recalling many years later that, "Philosophically, I liked [Hoyle's] steady-state theory of the universe except for the fact that it relied on untestable new physics." But the hot, big bang model, by contrast, made strongly testable predictions, as Penzias and Wilson were about to discover.

Penzias recruited Wilson in 1963. The two had rather different, but complimentary, personalities: Penzias was garrulous and interested in the big picture. Wilson was reserved, thought about the details, and liked to get his hands dirty by running the telescope. The two made a good team for the tricky job of using a horn receiver for making observations. Radio astronomers were then still in the days when the control of telescopes was manual. When Wilson arrived at Bell Labs in New Jersey, the Telstar satellite (the first active communications satellite for television transmission) had recently ceased to function, its electronics having been wrecked by the high-altitude nuclear test explosions of the Cold War. The chance destruction of Telstar liberated the 20-foot antenna for astronomical research.

Figure 4.6. Robert Wilson and Arno Penzias, with the Bell Lab Holmdel (NJ) radio receiver that detected the cosmic background radiation field—proving that the hot big bang theory was correct. 1978. (Lucent Technologies Bell Laboratories, courtesy AIP Emilio Segre Visual Archives)

The Bell Labs astronomers settled on the idea of looking for radiation from the vast spherical halo that had been proposed by Lyman Spitzer (the theoretical astrophysicist and plasma physicist at Princeton who initiated the programs to produce hydrogen fusion power on Earth) as a yet unseen component that must surround the Milky Way. To explain their observational technique, we will use an analogy from optical astronomy. The Sun is a strong source of radiation in the visible region of the spectrum, and its predominantly yellow light tells us that its surface, the photosphere, has a temperature of about 6000 K. In general, hot sources (stars, tungsten light bulbs, candle flames) emit visible light and infrared light. Cold sources of radiation do not produce visible light or infrared radiation, but they do emit in the radio part of the spectrum. Natural sources of the even longer wavelength, microwave radiation are still cooler and have temperatures of less than 30 K, some 240 K below the freezing point of water. A telescope observing microwave radiation would detect any cold sources, whether or not it was searching for them.

Penzias and Wilson were looking to detect the microwave radiation from the glowing halo of gas enclosing our galaxy. This required them to measure the temperature of the radiation detected by the antenna. However, they anticipated that the real temperature of the microwave, radio emission from the galaxy would be augmented, or contaminated, by radiation from the terrestrial atmosphere as well as from the antenna. Those terrestrial and instrumental contributions needed to be subtracted from the signal in order to measure the galactic component. That seemed to be an elementary matter because the necessary calibrations had already been made for communication purposes. However, when they turned on the telescope and looked out into space, the sky was, mysteriously, too bright in microwaves to see in it the clouds of gas for which they were searching.

The two young astronomers pressed on with observations. Gallons of liquid helium were on hand to provide a reference temperature of 4.2 K. They eagerly examined the results of the first observing run, poring over the shaky curve of a pen-recording voltmeter. It was immediately clear that something was seriously amiss. Or was it? The antenna, gazing at the sky, registered a temperature 3 K warmer than the liquid helium reference. An error of that size meant that the galactic halo experiment would not work, which initially was deeply disappointing for the pair.

In a scientific investigation of this kind, the first thing to check is that all of the sources of contamination have been accounted for correctly. The two asked if heat radiation from New York City might be the problem. However, when the antenna swept across the distant Manhattan skyline, nothing extra was seen. Radio sources in the Milky Way were not strong enough to be causing the temperature enhancement. That left the walls of the antenna. In a bizarre example of attention to detail, Penzias and Wilson even evicted a pair of pigeons that roosted in the antenna, and scraped away their droppings. This house cleaning made little difference. A year went by. Wherever in the sky they pointed the antenna, the recorded temperature remained stubbornly high, which led to a good deal of head scratching.

From an academic point of view, a disadvantage of the working arrangements at Bell Labs was that the researchers were relatively isolated within a corporate telecommunications environment. In a university or an astronomical observatory they would have interacted on a daily basis with colleagues having cognate interests. Penzias and Wilson were, in effect, cut off from recent developments elsewhere in astronomy that had not yet been published in astronomical journals. They even decided to publish their result on excess temperature as a section buried in another paper they were working on. Fortunately for the pair, fate intervened before they submitted the paper in which they reported the excess temperature as a puzzling aside.

In December 1964, Penzias attended the meeting of the American Astronomical Society in Montreal, where he chatted about their puzzling result with Bernie Burke, an experienced radio astronomer from MIT. Burke had no immediate suggestions for the cause, but he did not forget the conversation. Thus it was that a few weeks later Burke called Penzias with the interesting news that Jim Peebles at Princeton was predicting a temperature of about 10 K for the big bang fossil radiation. A few days later, Penzias received by mail a preprint of Peebles' paper. Penzias was of course delighted to find a possible explanation for the excess temperature, but he had little interest in the implications for cosmology. The *Physical Review* rejected Peebles' paper on the grounds that Gamow, Alpher, and Hermann had made a similar prediction back in the 1940s. That was news to Peebles.

Following the tip-off from Burke, Penzias called Princeton and spoke to Bob Dicke, Peebles' senior advisor, who was running a small experiment to detect the hypothesized background radiation. On hearing the startling news that Bell Labs had detected a temperature excess coming from the sky, scientists from Dicke's lab raced over to Holmdel, where they approved of the experimental setup, and pronounced the temperature excess a detection of the long sought cosmic microwave background with a temperature of 3 K. The Princeton group had been looking at the sky with a separate radio telescope, but their instrument was too small to reliably detect the feeble signal that the Bell Labs group had found by accident.

Both Bell Labs and Princeton published papers, back to back, in the issue of *Astrophysical Journal* dated July 1, 1965. On May 21, the *New York Times* ran a front-page scoop headlined "Signals imply a big-bang universe." Walter Sullivan, then the dean of science writers in the United States, apparently had a mole in the editorial office of the *Astrophysical Journal*. Sullivan's coverage noted that the discovery seemed incompatible with Hoyle's steady state theory. Sullivan wrote: "Dr Dicke clearly doubts the steady-state theory in which there is no explosion at all."

■ Understanding the Big Bang

The history of the discovery of the cosmic microwave background and its interpretation as the fossil radiation from the Big Bang is a good example of the twisting path that progress in science can take. As we have already noted, Lemaître was the first to contemplate a hot beginning for the expanding universe. Writing in *Monthly Notices* in 1931, he stated:

> The evolution of the world can be compared to a display of fireworks that has just ended: some few red wisps, ashes and smoke. Standing on a well-chilled cinder, we see the slow fading of the suns, and we try to recall the vanished brilliance of the origin of the worlds.

This quote is clearly a description of the hot big bang model, but Lemaître's candidate for the remnant radiation was cosmic rays rather than thermal radiation in the microwave band.

The search for the origin of the light elements had led to the suggestion that all of the matter in the universe had passed through a hot (10^{10} K), dense phase, in thermal contact with an extremely hot (X-ray) radiation field. Gamow, in 1948, had made an order of magnitude calculation showing that the present radiation temperature of our universe should be several degrees above absolute zero. In the 1950s and 1960s, Gamow's hot universe picture was known but not discussed. It received no attention from the U.S. astronomical community for which cosmology was all and only about tracking the expansion of the universe using optically observed galaxies.

Yet, in Moscow there was another school of investigators, led by the extraordinary polymath, Yacob Borisevich Zeldovich. Two of his younger colleagues, Andrei Doroshkevich and Igor Novikov wrote a paper (in 1963) saying that long wavelength radio observations should be able to detect Gamow's (1948) predicted radiation. This paper is the first one to emphasize that the spectrum to be seen at our epoch should resemble that in a "black body" as predicted by Max Planck in 1900. The Moscow paper predates the discovery. No cosmologists in western countries were aware of the paper, and none quoted it. But later in that year Zeldovich himself computed the expected temperature to be relatively high and argued against Gamow's theory. The Moscow group was working actively and independently on the subject of the expected background radiation field and ultimately made some of the most essential and original contributions.

Among most physicists in the mid-1960s, the action had switched from the origin of the universe to the origin of the chem-

Figure 4.7. Yacov Borisevich Zeldovich, versatile Soviet physicist who (like Bethe and others) made important contributions both to the weapons programs and to cosmology. (AIP Emilio Segre Visual Archives, Physics Today Collection)

ical elements by nucleosynthesis in stars. But, Bob Dicke's group at Princeton, and the Moscow group around Zeldovich, independently recognized that detection of blackbody radiation across the whole of the sky would provide direct evidence for the Big Bang. The Moscow theorists, like Friedman and Lemaître before them, did not publish in English, which undoubtedly led to their work being overlooked. To find the predicted radiation, Dicke had constructed a radiometer (a sensitive instrument to detect radio waves). He was unaware of the anomalies found by the much larger radiometer being used thirty miles away, at Bell Labs. By itself, those observations of unexplained excess thermal radiation did not constitute an important discovery. It was simply a curiosity that had been found. Until the intervention of the theorists from Princeton, Penzias and Wilson had merely observed something they could not explain, which happens repeatedly in astronomy. Nor were they that focused on cosmology. Their curious result became a discovery when interpreted by the Princeton physicists, and by the later experimental work confirming and extending their results after 1965.

The scientific advances we have described led to Nobel Prizes in physics for some of the researchers we have mentioned. In 1974, Ryle and Hewish were the first physicists working in astronomy or cosmology to receive the award. And in 1978, Penzias and Wilson were also recognized with Nobel Prizes. But the tale is tangled. As we noted earlier, nuclear physicists had by this time solved the problem of the creation of the heavy elements (carbon, oxygen, iron, and all the rest), demonstrating conclusively that they were made in stars and a good fraction of them in stellar explosions. William Fowler, Margaret Burbidge, Geoffrey Burbidge, and Fred Hoyle (B^2FH) published the classic work on the subject, "Synthesis of the Elements in Stars," in 1957 (see fig. 4.4). Then in 1983, William Fowler was awarded a share in the physics prize for this work along with Subrahmanyan Chandrasekhar, who had shown that there was a maximum mass possible for certain, condensed stars. Some professionals saw the omission of Hoyle's name in 1983 as an extraordinary slight. It may have been a reaction to Hoyle's earlier outbursts or simply due to the rules of the prize committee limiting the awardees to three individuals, so all of the authors of the B^2FH paper and Chandrasekhar could not have shared the 1983 prize.

As we have seen in retrospect, much of the impetus that led to the discovery of the cosmic background radiation arose from the classic, curiosity-driven question, probed from the pre-Socratic philosophers through the alchemists: what is the origin of the chemical elements? But there was another driver of inquiry, both more practical and more dramatic. In 1939, Yacob Zeldovich in Russia, Otto Hahn in Berlin, and Frédéric and Irène Joliot Curie in Paris independently wrote papers on nuclear physics, saying in effect that a chain reaction decay of uranium might be possible and that such a fission reaction would release a great deal of energy. The Joliot-Curies had successfully repeated the startling chain reaction experiment of Hahn's, and their paper had the biggest impact because they published in *Nature*.

Within a few years there was active war work on both sides of the Atlantic that led to fission-based atomic bombs and culminated in the U.S. and Soviet hydrogen bomb detonations in the 1950s. These stupendous explosions demonstrated in horrifying splendor just how much energy could be released by the fusion of hydrogen isotopes into helium. Hiroshima and Nagasaki showed the effects of fission bombs on civilian populations in wartime, but fortunately, the far, far greater force of the fusion bombs, which literally mimicked cosmic explosions, was demonstrated only in uninhabited areas. No sentient being on the planet was left in doubt about the awesome potential of nuclear explosions. While one cannot prove that scientists working in this environment were motivated to look for and to find the cosmic big bang, it cannot have been irrelevant.

So, to conclude the saga of the hot big bang model, it is clear that the pioneering investigations proved the existence of the cosmic background radiation and thereby amply demonstrated the reality of the hot big bang. We were seeing the afterglow directly in the sky as well as in the ashes in the light elements produced in the Big Bang.

The early microwave experiments had poor accuracy and inadequate angular resolution. If there was structure in this cosmic background radiation, indicated by variations in the level of radiation from different parts of the sky, it was undetectable at that time. Soon it was realized, however, that there *must* have been early, low-level fluctuations from place to place in the distant sky or else the local universe would not be as complex as it is. The most obvious and unsophisticated observation that one can make (that the universe does have structure) had deep implications. As we shall see in the next chapter, detailed scrutiny of the cosmic microwave background fluctuations was to be immensely important for understanding the origin of structure in the universe.

Chapter Five

The Origin of Structure in the Universe

■ "In the Beginning"—Why an Explanation Is Needed

Until recently, astronomers had never asked the obvious question that philosophers had often asked. The German existentialist Martin Heidegger rated the question "Why is there something rather than nothing" as the most fundamental issue in philosophy. Why does existence exist? All attempts to unravel the evolution of existence become an infinite regression that may be, in fact, a philosophical dead end. However, a cosmological variant of this question had become ever more pressing after cosmologists realized in the 1960s that they should be able to answer it: "Why is there structure in the universe and from what does it arise?" This conundrum is central to our history of cosmology, and the search for the answer became a prime mover in leading us to the discovery of dark matter and dark energy.

The first book of the bible, Genesis, starts out with "In the beginning, the earth was without form and void," and this reasonable start is one that present-day physicists would be inclined to adopt as well. To a certain extent it is empirically true, since we find that on the largest observable scales the cosmos is fairly smooth, the same in all directions, and with decreasing irregularities, as we look farther away in space and further and further back in time. But, after a moment's more thought about the matter, we can see that there is something not quite right here. If the cosmos had started out perfectly smooth, then, without a god to stir things up, would it not have stayed perfectly smooth? And then, whence came all the structures that we see, the galaxies, stars, and planets? This is the major issue, explaining how structure arises in an apparently featureless universe. Scientists only realized that they had to address this fundamental question, somehow or other, two-thirds of the way through the twentieth century.

The prize-winning discovery of the microwave background in 1965 consigned the steady state model of cosmology to oblivion in the public eye. In fact, the evidence against this appealing but erroneous theory had been compelling to all but enthusiasts for some time. Now, the radiation from the hot big bang had been seen. Cosmologists began to consider afresh the consequences for a universe that evolved from a hot, dense, nearly uniform plasma of light and particles to its current heterogeneous mix of isolated astronomical objects. Cosmology became the arena in which questions about cosmic evolution and the origin of structure could be debated. Whatever picture cosmologists developed for the mature stages of the universe, as seen with our optical telescopes focused on our environs, needed to be consistent with an origin in a hot big bang start.

The first generation of cosmologists had taken galaxies, which were the evident building blocks of the visible universe at our epoch, as being simply "out there"—objects to be used as markers of the expanding universe. But gradually a consensus developed that good cosmological models *must* be able to account for the formation of galaxies and their observed distribution in space. The central question became: How does structure arise in an expanding universe? If the universe had been perfectly uniform, it would have stayed uniform; galaxies, stars, and planets would never have appeared. But, if it had begun with small fluctuations embedded within it, then there must have been some powerful mechanism causing them to grow, to produce what we see in the visible universe at our epoch. On the largest scales Newton's and Einstein's gravity is the dominant attractive force, so it was natural to look to it for the genesis of the observed structures

The British cosmologist, James Jeans, was the first astronomer to assess quantitatively how gravitational instabilities could cause structures to grow from very small initial fluctuations. In graduate schools today, the process is still called "the Jeans instability." He had graduated from Cambridge in 1900 with highest honors, for which Trinity College elected him a Fellow. Within two years, twenty-five-year-old James Jeans made a major discovery about the physical conditions that must be satisfied for the formation of structure in the universe.

The chapel at Trinity College has a striking statue of Isaac Newton; all Trinity men knew that Newton was (and still is) their most famous alumnus. So it is not surprising that, on appointment to a fellowship at Trinity, Jeans decided to apply Newton's law of grav-

Figure 5.1. James Jeans, presidential portrait. (Royal Astronomical Society)

ity to the cosmic conundrum of how nebulae form and condense from gaseous matter in space. Jeans was also strongly influenced by Sir George Darwin (the second son of Charles Darwin), who held the Plumian Professorship of Astronomy at Cambridge and who studied the tidal forces involving the Sun, Earth, and Moon; the problems encountered in these studies are closely related to those of explaining the gravitational origin of heavenly bodies.

Jeans showed how small perturbations could grow into large ones due to gravitational forces alone. The Jeans instability is still the best explanation for the collapse of interstellar molecular clouds, the precursors of forming stars. Jeans had made this remarkable discovery: an interstellar cloud will be stable and not collapse if it has a sufficiently small mass, but once a certain critical mass is exceeded, the cloud will begin a runaway collapse. The critical mass is now named the Jeans mass. As a means of making galaxies, the Jeans mechanism is a good starting point.

So far as star formation inside a galaxy is concerned, the Jeans theory has held up reasonably well despite some technical flaws that theorists later found in the analysis. But Jeans was thinking of stars, not galaxies. Can the physical conditions envisioned by Jeans be scaled up to apply to the formation of galaxies themselves? A galaxy is a stable, gravitationally bound system. Theories that successfully explain fragmentation and structure formation *within* a galaxy might be applied to the extragalactic realm to the universe as whole, but only if the universe itself were a stable, static structure. However, as we have seen, by the 1930s it had become clear, after an uphill battle with Einstein in the previous decade, that the universe is expanding rather than stationary, so Jeans' analysis, while useful as a starting point, could not answer the question of how structure would grow in an expanding universe. The physical reason for the difficulty is simple; the Hubble expansion velocity is proportional to distance. Therefore, in widely separated pieces of the universe, the kinetic energy of expansion (which is proportional to the square of the velocity) can easily overwhelm gravity. This will limit the size of pieces that can fragment out, so a more careful treatment is needed.

Cambridge became the center of activity. Jeans moved back there in 1910 after a brief stint as professor of applied mathematics at Princeton University. Eddington, who had departed Cambridge in 1906 for the Royal Observatory, returned in 1913 as Plumian Professor of astronomy (a post later held by one of the authors of this book). During World War I, Jeans and Eddington were both in Cambridge and sometimes dined together at the Trinity high table. Jeans was too old for compulsory military service, and Eddington was, as we noted, a determined pacifist who did not enlist. It was during the war years that Jeans experienced his finest period as an astrophysicist, working on the collapse through gravity of rotating masses.

No scientific progress from Jeans' fundamental analysis was made for half a century, since it took that long for the comprehensive model to be established within which galaxy formation would occur. Our understanding of the growth of structure in the universe came together slowly. Observations of different phenomena, some seen by optical telescopes, and some observed in the radio, had to be combined into a coherent picture having a consistent historical development. How could a nearly featureless, early quantum soup evolve to the present, complicated, real world, containing manifold structures visible at every physical scale?

Several strands of the answer to this puzzle grew independently, cohering only by the end of the 1980s. The observational strand of the process, using ground-based optical telescopes, was dramatic and culminated in the photographic and redshift surveys of galaxies by Margaret Geller and John Huchra (1989) at Harvard, who found fascinating and unanticipated structures, such as "walls," "voids," and even "stickmen" in the large-scale distribution of galaxies. But what caused the formation and evolution of the building blocks, the galaxies, themselves? In 1977, Martin Rees at Cambridge and one of the authors (Ostriker) showed how there is a mechanism similar to that discovered by Jeans and a critical cosmic mass (of approximately one hundred billion suns) that characterizes the formation of massive galaxies. The mechanism seems to set the characteristic or critical mass for galaxies—the mass to

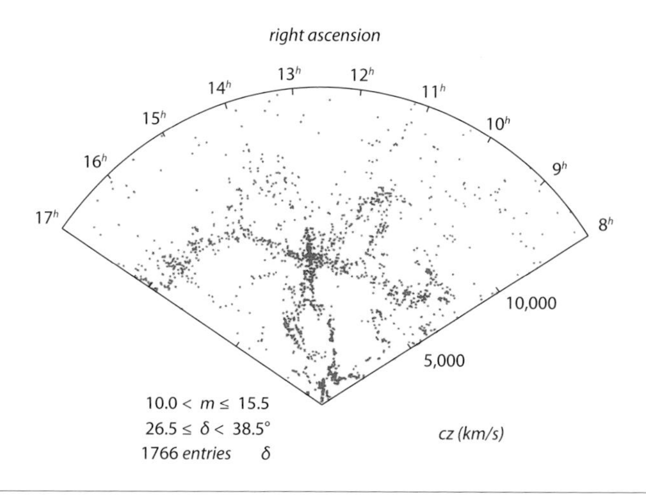

Figure 5.2. In this cone diagram the observer is at the vertex looking out into space and each dot is the position of a galaxy (in angle and distance from us as given by its redshift in km/s). It was the first to show large-scale structure in the universe (1989). The "stickman" is a cluster of galaxies embedded in the "great wall" of galaxies. (Work by M. Ramella, M. Geller, and J. Huchra (*Astrophysical Journal*, 344: 57, 1989)

which galaxies seem to grow by natural gravitational processes. Several other scientists wrote similar papers at that time, and it does indeed seem as if there is a "natural" scale to which galaxies will grow.

Going back still earlier in time, there was the question of what initiated the process of galaxy formation. A separate strand of investigation was to find in the sky the initial seeds from which these galactic structures started. And, finally, we needed to discover the origin of the seeds and the physical processes by which they grew to form the forest around us. It is a fascinating and winding tale that unfolded more than fifty years ago and will be told in this and succeeding chapters.

But, before we embark on the journey, let us review what exactly were those structures seen in the sky whose origins begged for explanation?

■ Structure within the Expanding Universe

Modern measurements of the microwave background show that the universe began in a hot, dense, nearly uniform state, approximately 13.7 billion years ago. But the sky today is far from uniform. We see structures on all scales, from planets and stars, to galaxies and, on much larger scales, galaxy clusters, and enormous voids between galaxies and even cosmic scale "filaments" and "great walls" of galaxies (see fig. 5.2). The famous Hubble Deep Field (see fig. 5.3) images taken by the Hubble Space Telescope provide stunning evidence of the structural variety in the universe. A good cosmological theory must provide an explanation for the origin of that structure, and it must be able to show how small seed perturbations in the hot plasma of the early universe could have grown to what we see today.

Let us start within own neighborhood. We have, as noted, a big brother galaxy, Andromeda, who is (somewhat ominously) approaching us rather than receding from us. There are many other smaller members of the local group. The whole comprises perhaps fifty galaxies, is about three million parsecs in size, and seems to be an approximately stable grouping. (The astronomical unit of the parsec is the distance an observer would be from our solar system to see our orbit around the Sun as being one arcsecond in angle on the sky. One parsec is 3.26 times farther than the distance that light can travel in a year, and roughly equal to 30 trillion kilometers.)

Outside of our own local group there is a concentration of galaxies in the direction of Virgo. About 20 million parsecs in extent, this high-density region subtends an arc of roughly eight degrees on the sky. Our local group of galaxies is considered to live in the outskirts of the Virgo Supercluster.

But, as we look farther out, at distances greater than 100 million parsecs (100 Mpc), the universe is beginning to look rather smooth, having patches of higher density scattered about with a separation of roughly 50 Mpc. On this scale we are beginning to see the "uniform cosmos" postulated by the theoreticians. The velocities do satisfy the Hubble law of uniform expansion. The additional small

Figure 5.3. The Hubble Ultra-Deep Field, roughly a quarter the size of the Moon on the sky. The deep view, taken in late August 2009, provides insights into how galaxies grew in their formative years early in the universe's history. The faintest galaxies seen formed 600 million years after the Big Bang. (NASA, ESA, G. Illingworth, UCO/Lick Observatory and the University of California, Santa Cruz; R. Bouwens, UCO/Lick Observatory and Leiden University; and the HUDF09 Team)

peculiar velocities, due to the motion of galaxies within groups, are irrelevant compared to the overall expansion. On this larger scale, if we even out the fluctuations, we do see the simple model that Hubble, Friedman, Lemaître, and others had analyzed with the aid of Einstein's equations of relativity. However, on smaller scales, there is an incredible amount of detailed structure, which we had, so to speak, stepped over in our eagerness to address the big picture, and it was only in the 1960s and 1970s that cosmologists realized that it was incumbent on them to provide some explanation for it.

Why is the formation of galaxies in an expanding universe an important intellectual puzzle? In general terms there are three processes to consider. The fact that the universe is expanding means it is literally flying apart, which disperses the matter in it. A second process tending to push matter apart is the thermal motion or heat energy of the matter. The thermal opposition to collapse and fragmentation is what Jeans analyzed at Cambridge. So the universal expansion and the thermal energy conspire to push the universe in the direction of a featureless continuum with no structure. Acting against this is the universal force of gravitation, which not only will keep matter together, but will tend to concentrate it into ever more dense lumps. Which processes win and under what circumstances one or another will dominate are nontrivial questions that physics should be able to answer.

In some astronomical situations the forces can come into a balance. An ordinary star like the Sun is a good example. Clearly the Sun is stable or else we would not exist. We say the Sun is in hydrostatic equilibrium; this technical term says that the gravitational force trying to crush the Sun is exactly balanced always and everywhere within the solar interior by the pressure released from the heat energy generated in its core. But in the universe this balance did not happen; the state of expansion was not in equilibrium. The early universe was unstable, but somehow small regions of local high density nevertheless managed to clump into galaxies before they could be dispersed. The problem comes down to a race against time. Cosmologists working in the Big Bang paradigm needed to find the Goldilocks solution in which the early universe possessed the right temperature, the right density, and the right expansion rate, so that gravity could assemble the first galaxies before the universe flew irrevocably apart.

When Gamow first proposed what later would become known as the hot big bang concept, the difficulties of understanding galaxy formation in a rapidly expanding universe stood in the way of wider acceptance of the big bang model. The more one thinks about this problem, the more puzzling it becomes. If the initial perturbations had been too small, the universe would have stayed uniform (and we would not be here to ask why that happened). But, if it had started out too lumpy, then great pieces would have collapsed into cosmic-sized black holes at early times, and we probably also would never have come into being. How and why did it transpire that things were just right at early times? That question touches on the "anthropic principle," a philosophical construct based on the fact that the physical parameters of the universe happen to be compatible with the existence of intelligent observers. Later in this chapter we will ask whether the fascinating anthropic principle is profound or empty.

Whatever the answer to this question (and the jury is still out), it seems that conditions did favor the creation of galaxies. Roughly 10 percent of the ordinary matter in the universe has collapsed into these rather gorgeous stellar assemblages that, individually, are not expanding but seem to maintain a serene internal equilibrium in an evolving universe. These stellar systems obviously could not have existed at early times in a hot big bang universe. They must have been formed over time from growing irregularities, as the gas and radiation from the Big Bang expanded and then cooled. How this could have happened was a rather obvious intellectual conundrum, but somehow the savants looked the other way for decades and avoided the difficult problem. Historically they were content to see galaxies as the building blocks of the universe, of unspecified and unexamined origin and evolution until they were absolutely forced to consider their genesis. The issue that rubbed their noses into this messy subject was the need of the observational cosmologists to have good standard candles, objects of fixed luminosity, to use in their studies of the expanding universe.

■ The Elusive Standard Candle: Beatrice Tinsley Changes the Game

The goal of cosmology and most extragalactic optical astronomy during the heroic period spanning the half century from Hubble to Sandage (1920s–1970s) had been to simplify the subject of cosmology to the measurement of just two numbers: the Hubble constant, H_0, which stood for the current rate of expansion of the universe, and q_0, which was known for decades as the "deceleration parameter," the value of which would tell us whether the universe would expand forever (q_0 less than ½) or re-contract and then collapse (q_0 greater than ½) ; see appendix 1 for details. As we showed in our chapter 3 tour of early observational cosmology, this was the focus, the holy grail, of scientists worldwide: to obtain the two numbers that would tell us which of the possible world models was the real universe in which we lived. The great Palomar 200-inch telescope, finally completed in 1949, had been built specifically to answer these global questions. In passing, we note that we now know that just two numbers are insufficient to specify the universe: there is an extra component, dark energy, that must also be measured, and this strange stuff causes the universe to accelerate. We tackle this in chapter 7, while noting that Hubble and Sandage were not aware of dark energy.

To do geometry, whether in the classroom or the universe, one needs some kind of standard objects, such as standard meter sticks. For this purpose the astronomers wanted to be able to use galaxies, the actual, real objects seen in the extragalactic world. This was the era when the two authors of this book were students and young researchers. Although it seems odd today, both observers and theorists, during this extended period, used galaxies primarily as standard candles and standard meter sticks, and ignored the issue of their origins and evolution.

However, the investigation to determine H_0 and q_0, using galaxies as standard candles or measuring sticks, required certain assumptions. It was obvious that, if all galaxies were of the same brightness, then observations of the apparent brightness as a function of observed redshift (or distance) could help to determine q_0. How does that work? In a simple Euclidian universe the apparent brightness would fall off as the square of the distance from us. Deviations from this simple scaling law would help to measure q_0, and tell us about the global geometry of the universe. Also, if they were all of the same physical size, then seeing their apparent or angular size would tell us how far away they were.

But this would work *only* if all galaxies had identical properties, which is manifestly not the case. Detailed studies of the 1950s to the early 1990s attempted to calibrate these systems and to learn how their luminosities and sizes could be correlated with each other and with other observable qualities like color and shape. Then, even if they were not identical, we would have enough information to correct for the natural variation among the sample of galaxies. But the assumption remained that, whatever properties galaxies had at our epoch, they were unchanging; there was no need to allow for galactic evolution.

The huge effort to calibrate the galaxies so that they could be used as standard candles to measure the universe was undermined, in a way that no cosmologist had anticipated, by a graduate student, Beatrice Tinsley, who brought the great enterprise crashing down when she noted the obvious: that ordinary galaxies *must* be changing with time and could not be treated as standard candles!

Tinsley was born in January 1941 in war-torn England, her parents relying on the generosity of neighbors to share their meager rations of coal. In 1946, the family emigrated to New Zealand to start a new life. Young Beetle, as she was nicknamed, excelled at school, and she was an outstanding musician who could have had a professional career. For two years she performed with the New Zealand National Orchestra. However, at the age of fourteen the rising star had announced her ambition to become an astrophysicist. She majored in physics at the University of Canterbury, receiving her degree in 1961. In December 1963, Tinsley attended the first Texas Symposium on relativistic astrophysics. She wrote, "the people coming include Hoyle, Oppenheimer and just about every famous living man in the field from all over the world." It did not take her long to find her place among these luminaries.

She enrolled for a PhD program at the astronomy department of the University of Texas at Austin, which was a five-hour commute by public bus from her home. Her degree was awarded in 1966, after only two years of study—the speed of completion show-

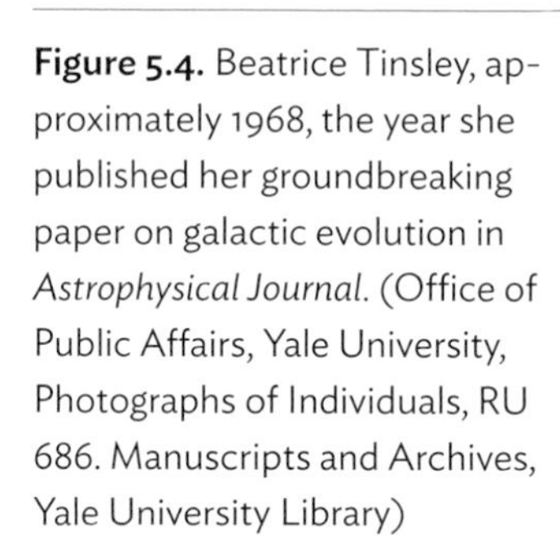

Figure 5.4. Beatrice Tinsley, approximately 1968, the year she published her groundbreaking paper on galactic evolution in *Astrophysical Journal*. (Office of Public Affairs, Yale University, Photographs of Individuals, RU 686. Manuscripts and Archives, Yale University Library)

ing her extraordinary ability. The title of her thesis was *Evolution of Galaxies and its Significance for Cosmology*. Her doctoral research was highly significant; it was the first step in showing that the great enterprise of calibrating the galaxies was in need of major revision. It should have been obvious that, since galaxies were not existent in the beginning, they must have formed over time, and thus they must be evolving. They could not be used as standard objects of measurement until the process of formation and evolution was understood or at least estimated.

Immediately upon completing her PhD, Tinsley used her tenure as a University of Texas Fellow to launch a more thorough investigation of the evolution of galaxies. She was just twenty-five years old when she began, working in a somewhat isolated environment, but she was strongly supported by her husband who also worked in astrophysics. Tinsley computed how galaxies would change in appearance over history due to the calculable evolution of their constituent stars. She computed the evolution, beginning with star formation from interstellar gas. As her classic 1968 paper in *Astrophysical Journal* puts it, she made "a machine-computed evolutionary history of a galaxy" from initial formation to an age of 12 billion years. As an aside, we note that her programming code was based on Fortran and ran on a machine with 16k of magnetic core memory; in other words, she used a state-of-the-art scientific computer but with a power less than that in a talking greeting card!

The basis for Tinsley's conclusion was simple and not disputable; normal stars gradually use up their fuel and evolve with time. The evolution of the Sun had been understood since the late 1950s and, if it must use up its fuel, evolve, and ultimately die, so must the similar stars in the galaxies that were being used as standard candles by Sandage and others. The final paragraph of her paper gave this chilling warning to the cosmologists:

> The application of the computed past history of a [giant elliptical] system to cosmology has shown that evolution may be of great importance to the interpretation of [the magnitude-redshift] relations.

The citation history of this paper is revealing. At the Mount Wilson and Palomar Observatories, John ("Bev") Oke and Allan Sandage responded quickly. The first sentence of their 1968 paper is: "The most pressing current problem in observational cosmology is the determination of the Hubble constant H_0 and the deceleration parameter q_0." Their final sentence dismisses Tinsley as being too pessimistic. The Oxford astrophysicist John Peach entered the fray in 1970, beginning his paper on redshift-magnitude relation by repeating the obligatory sentence on the most pressing problem in observational cosmology. He cites Tinsley's models as differing considerably from Sandage's, and then pronounces that he "will accept the most recent calculation of Sandage."

Undeterred, Tinsley began to promote her results more actively. In June 1970, she had a paper at the 132nd Meeting of the American Astronomical Society, held in Boulder, Colorado. We have only found the abstract of this paper, in which she boldly wrote:

> Galactic evolution may be rapid enough to affect significantly comparison of the observed magnitude-redshift rela-

> tion with cosmological models. [My] present conclusion is opposite to that reached by most cosmologists.

In August 1972, Tinsley and her colleague Jon-Erik Solheim presented a paper on the effects of evolution at a symposium on galaxies and quasars held in Uppsala, Sweden. The concluding paragraph of their paper began as follows:

> The results show that evolution could have considerable effect on the determination of a cosmological model from the magnitude-redshift relation.

Beatrice published eleven papers in 1972. Toward the end of that year she finished a pair of classics on the evolution of giant elliptical galaxies published in *Astrophysical Journal.* The longer paper, published on December 1, is relatively understated with respect to the cosmological implications, although she did permit herself to note that observations of the kind used by Sandage (and others) would over-estimate the value of q_0. In a short *Letter*, published two weeks earlier, she explicitly challenged the (then) standard assumption that galaxies had a fixed physical size for determining the cosmological parameters, and effectively undercut Sandage's enterprise, but in a typically polite manner.

Tinsley's insistence that galaxies *must* evolve with time caused great consternation in the community of classic cosmologists. The level of alarm was almost palpable; what had this young woman done? There were unsubstantiated rumors that there had been efforts to suppress the publication of her papers, to suppress the beast of doubt before it could leave its cradle. Huge amounts of money, prestige, and time had been invested in the classic enterprise of cosmology, to measure just two numbers. But, in fact, the scientific system worked as it should have worked. Tinsley's papers were refereed, published, and soon acknowledged as being of central importance. Her paper published on November 15, 1972 was in print after just ten weeks.

In this fashion, the scientific community was, so to speak, backed into an acknowledgment of ignorance and subsequent analysis of the origins of galaxies and in fact all cosmic structure—compelled by the desire to use galaxies as tools in cosmological investigations. Logically, quite apart from any practical utility in such understanding, its absence showed a gaping hole in our conceptual framework.

It should have been obvious to all that, if the universe started out in a *truly* homogeneous state, containing no density or velocity fluctuations, then it would have stayed in that uniform state. In a perfectly uniform universe there could be no net force on any little element of mass—because there is no preferred direction and therefore no direction in which the force could be pulling the matter. Clearly this was not the initial state. The very existence of galaxies indicated that fluctuations were intrinsic and fundamental to cosmology, not an add-on, and must be faced as an *ab initio* element of the model. The version of the start of the world given in Genesis needs, in a physics-based model, the extra ingredient of initial perturbations; galaxies were the obvious consequences of such initial fluctuations.

If galaxies were the bricks in the buildings of the universe, it was now realized that a brick factory was needed, and the modern enterprise of studying galaxy formation was initiated. This proved to be a rather heavy computational task, and we will have more to say about it in our final chapter. Now, however, based on these calculations, we can respond to journalists' questions of where the galaxies were formed, when they were formed, and how they were formed; we are now just beginning to confront the much harder questions concerning their detailed interior structures.

Let us follow the common metaphor of galaxies as the building blocks or bricks of the universe. These bricks are arrayed into buildings (groups of galaxies) and those are organized along streets (filaments) with some fraction found in giant cities (massive clusters of galaxies containing thousands of objects). The simple picture of a uniform distribution of galaxies is too simple, and a whole field of science was developed to find, catalog, and analyze the large-scale structure of the universe.

■ Real Cosmic Structure Found and Cataloged by Fritz Zwicky

The observational quest to understand the structures found in the universe had, of course, preceded the "quest for two numbers." From the time of William Herschel (1780s) onward, the observers had asked if evidence of structure could be discovered telescopically, and Herschel tried to map the features of the galaxy in three dimensions. By the late nineteenth century, astronomers were using photography to reveal structure within what were termed the white nebulae (but we know today as galaxies). In the 1880s, a gifted amateur, Isaac Roberts, exhibited stunning photographs of nebulae at the monthly meetings of the Royal Astronomical Society and was the first to produce a photograph of our friend, the Andromeda Nebula, M31, that is instantly recognizable to this day, a half-degree in size, similar to the Moon, and visible to the naked eye in the winter sky. The deep images that Roberts took were the first to demonstrate the power of photography in revealing the heretofore hidden depths of the universe.

By the turn of the century, at the Lick Observatory, California, director James Keeler was producing exquisite photographs, using the 36-inch Crossley reflector. Importantly for our story, Keeler found vast numbers of undiscovered nebulae on his plates, leading him to predict that 120,000 nebulae were within range of his 36-inch telescope. Indeed it is intriguing to imagine, but not historically correct to do so, what Keeler might have achieved with the 60-inch at Mount Wilson but for his untimely death in 1900.

Photographic sky surveys became established partly as a result of the efforts of Fritz Zwicky, the Swiss-born astronomer of extraordinary originality who worked at Caltech. We have noted him as a brilliant maverick, many of whose speculations were so far ahead of their time that he was regarded as a crank; and we relate more of his career in the next chapter. Even in a subject that many would consider fairly dry, the cataloging of galaxies, Zwicky injected controversy by insulting his colleagues in Pasadena. We do not need gossip: in a self-published catalogue of galaxies of 1971 he writes:

> E. P. Hubble, W. Baade and the sycophants among their young assistants were thus in a position to doctor their observational data, to hide their shortcomings and to make the majority of the astronomers accept and believe in some of their most prejudicial and erroneous presentations and interpretations of facts . . .
>
> Today's sycophants and plain thieves seem to be free, in American Astronomy in particular, to appropriate discoveries and inventions made by lone wolves and non-conformists, for whom there is never any appeal to the hierarchies and for whom even the public Press is closed, because of censoring committees within the scientific institutions.

It is striking to note how tolerant the scientific community was (and is) of idiosyncratic personalities. His evident brilliance was such that, despite his quite difficult character, Zwicky made it to the very top of the tree, to be a tenured professor at the foremost observatory on the planet. Perhaps the greatest of his achievements that was recognized during his lifetime was making photographic sky surveys. For this he has received too little credit. There is no Zwicky telescope, nor a Zwicky spacecraft, nor even a Zwicky parameter.

Fritz Zwicky had entered extragalactic research through an interest in cosmic rays, ultra-high-energy particles from outer space that collide with the Earth's atmosphere. Today we know that they originate from a variety of sources: those of lowest energy are from the Sun, whereas the highest energy cosmic rays reach us from the depths of the universe and their origin remains unknown. But when Zwicky first heard about them, at Caltech, their origin was a subject of lively controversy. He speculated wildly that cosmic rays were spawned in stupendous explosions of super-massive stars. He teamed up with Walter Baade, one of the greatest observers of the twentieth century (whose other work we discussed in chapter 2). Together, Zwicky and Baade worked on stellar explosions. In 1931, Zwicky introduced the term "supernova" in a lecture course, and it is the term we still use for exploding stars. At a conference in 1933, Baade and Zwicky spoke of how a massive star would end its life in a supernova explosion that blew the star to bits, creating cosmic rays and leaving a dense neutron star—a remarkable type of star, approximately the mass of the Sun, but made almost entirely of neutrons—essentially a giant atomic nucleus, similar to those in ordinary atoms but 10^{56} times more massive. This remarkable prediction, of course, long predated the discovery of neutron stars.

Zwicky and Baade then persuaded the Mount Wilson Observatory to buy an 18-inch Schmidt telescope, invented in 1930 by Bernard Schmidt, with an optical design that allows it to accurately survey large patches of the sky, so that he could search for supernovae in nearby galaxies. The wide field telescope commenced operations in 1936, at Palomar Observatory. With it, Zwicky discovered a dozen supernovae, and demonstrated that the Schmidt was very well suited for sky surveys. The cosmological thrust behind the research was to see if distant supernovae could be used as standard candles to calibrate the faraway universe. That "supernova cosmology" was finally achieved fully only in 1998, the centenary of Zwicky's birth. And, when dark energy was crowned with a Nobel Prize in 2011 as majority owner of the global energy density, it was based on the work of two teams of scientists using supernovae as the guides and signposts in their mapping of the current universe.

Building on the supernova success, Zwicky used the 18-inch telescope to make the first photographic survey of the entire northern sky, mapping thousands of galaxies. The resulting database, of almost 30,000 compact galaxies, was published as a six-volume catalog. As a result of this research, Zwicky discovered that galaxies tended to gather in clusters, thereby opening up a new chapter in the search for structure in the universe. This was the first evidence for ensembles bigger than galaxies that were held together by gravity.

At Palomar the 48-inch Schmidt telescope came on-stream in 1948, and it produced the famous National Geographic Palomar Observatory Sky Survey (POSS), showing stars down to magnitude 22. If there had been any doubt about Zwicky's claims concerning the importance of clusters of galaxies (see figures 1.2 and 6.1) as fundamental building blocks in the universe, the POSS plates provided unambiguous evidence that most galaxies are in clusters of various sizes. When the survey was being made, the principal observers took great care to examine the plates taken each night. George Abell classified the rich clusters of galaxies, producing a catalog of 2,400 of the richest, symmetrical, clusters. By the 1970s, evidence was emerging that galaxies were distributed along stringy features outlining emptier cells on the largest scales, with clusters at the vertices where the strings intersected—the "cosmic web" (see figure 8.1).

Cosmologists began to speak of the structure as being "sponge-like," with evidence for "holes," "filaments," and "walls." However, the distribution maps based on photographic surveys were two-dimensional, apart from analyses of the nearby galaxies, where distance information was available from redshifts. The key question then became: does the spongy structure continue out to great distances? To answer that question, a large program of investigation to measure redshifts of galaxies commenced. In the period 1930–1970, astronomers used photographic sky surveys to discover the structural arrangements of galaxies and clusters of galaxies. This was structure on scales far larger than the sizes of individual galaxies. If understanding galaxy formation had been a difficult enterprise, then how much harder would it be to comprehend structures that were a thousand times larger?

■ Understanding the Origin of Structure Becomes Serious Science

The existence of stars, galaxies, and clusters of galaxies tells us something about the seed fluctuations with which the universe began. It took decades for this rather unnerving thought to settle in, and for cosmologists to grasp the fact that the most imposing aspect of the universe is its internal structure, rather than the specific values of the global cosmological parameters. The more astronomers thought about the issues, the more problematic they became. On a simple level it is evident that dense regions will tend to expand more slowly than average regions, as they have to fight stronger gravity, and so they will become relatively denser still. Conversely, under-dense regions will expand more rapidly than average regions and so become yet more diffuse. This is the working of the classical gravitational instability, the details of which were analyzed in the 1960s by two separate groups: the Moscow school of Yakov Zeldovich (see figure 4.7) and his students, including Rashid Sunyaev, and the Princeton school of James Peebles and his associates. Their calculations were basically an update of the half-century-old Jeans' results to allow for the cosmological framework.

Given this generalization, the need for fine-tuning became embarrassingly evident. Suppose that the initial fluctuations had been very, very small. Then, even by growing through gravitational instabilities, they would still be very small now and there would be no clusters, galaxies, stars, or planets. Conversely, if the fluctuations had been very large, then long ago they might have already collapsed into massive black holes. In some sense they had to be just right. This is what we have termed the "Goldilocks Problem." As soon as this was realized, the Moscow and Princeton groups independently undertook computations to see what amplitude the fluctuations must have had at the era of recombination (when the hot soup of protons and electrons cooled enough to combine into atoms of hydrogen), at approximately a redshift of 1000, roughly 300,000 years after the Big Bang (see the cloudy region between red and blue in figure 4.3, p. 110), in order to give us the structures that we see in the local universe today. This epoch, called recombination, was a special time. On average, any two atoms in the present universe were a factor of one thousand closer together than they are now: the distance they are apart now is roughly two meters, so it was two millimeters at that time. The cosmic microwave background radiation comes from just that epoch. Before the recombination era atoms did not exist. The universe back then comprised electrons, protons, and radiation, and this plasma was opaque to its own radiation, so our search into the past stops at that era: the universe before recombination does not send us electromagnetic messages. We cannot see that early universe.

When the microwave background was discovered, it was immediately realized that any maps of the intensity of the radiation would reveal, at some level, an imprint of the primordial structures. The more dense parts would be brighter and the less dense ones would be fainter. These brighter and fainter spots should be seen if a detailed sky map could be made from the radio observations of the background radiation. One could ask how big the fluctuations in density must have been at this early epoch in order to have grown into the giant structures that we see around us. At the time of the discovery of the background radiation, cosmologists were already fairly sure that galaxies and large-scale structure had separated out from the expansion of the universe at redshifts of less than 100 when the universe was at most 1/500 of its present age and their infant precursors should be visible (if faint) in the microwave maps. The argument used to reach this conclusion is relatively easy to follow. Quantitatively, it can be demonstrated that, if the gas from which galaxies were to form had an over-density of one part in ten thousand (0.01 percent), this would suffice to make galaxies fully formed by the present epoch. This seems like a small density contrast, but it's certainly not vanishingly small and we present an elementary quantitative argument in the next chapter.

The origin of the necessary density perturbations has to be assigned to processes in the very early universe. But we can separately ask *when* the galaxies crystallized out, and this can be estimated by noting that the present average density within galaxies should be similar in broad terms to the average density of the universe at the epoch when the galaxies froze out or coagulated as separate self-gravitating units. This simple point, that galaxies, when they form, must be some round number (roughly 150) times the density of the universe at that time was first shown by Martin Rees and codified in a paper by Simon White (then at Cambridge) and Rees in 1978. We now know that this condition applied for massive galaxies at a redshift of about five to ten when typical atoms were about one meter distant from their nearest neighbors.

A vigorous search then commenced for the expected fluctuations that *must* have existed to explain the origins of the abundant normal galaxies. It was argued that these fluctuations should become visible in the background radiation when the accuracy of the measurements reached a level of one part in ten thousand. Then the "seeds" from which later galaxies would grow would have been found.

To estimate the level of fluctuations to be expected one would have to know, quite precisely, at just what epoch one was observing when one looked at the background radiation field. Cosmologists could then model the temperature history of the universe, a task first tried by Alpher and Herman back in 1948. This logical step is quite easy to explain: the temperature of the background radiation scales with redshift as T_0 $(1 + z)$, where T_0 is the temperature we observe today (2.7 K), and z is the redshift. With this linear relation we can now do some simple physics. For example, for what value of the redshift, z, is the universe hot enough to ionize all the intergalactic atoms? Ionization is the process of stripping the electrons away from an atom. For hydrogen, the simplest and most common atom, the ionization process separates the atom into an electron and a proton. At a temperature of about 4000 K, photons have sufficient energy to break a hydrogen atom into its constituent proton and electron. An easy calculation shows that hydrogen would be completely ionized at redshifts above $z \sim 1500$ (corresponding to a temperature of 4000 K).

Now let's reverse this thinking by asking about the history of hydrogen and helium in the expanding universe. At the epoch matching a redshift of $z \sim 6000$, hydrogen is completely ionized and helium mostly so. The matter of the universe at this point therefore consists almost entirely of electrons, protons, and helium nuclei (note the red zone in fig 4.3, p. 110). As the universe expands and cools, the hydrogen will recombine into atoms at redshifts below ~1500. That is why this special time is referred to as the epoch of recombination. In fact, over the redshift range ~1500 to

~1000 (corresponding to the temperature falling from 4000 K to 3000 K), photons, electrons, and excited states of atoms are interacting vigorously.

Then, at a redshift of 1000, a crucially important event occurred: as we noted earlier, the expanding universe became transparent to its own radiation. Prior to this event, light of any kind was being scattered by free electrons before it could travel very far. In effect the electrons acted like fog. But after those electrons combined with protons to form hydrogen, photons could suddenly travel unhindered, and they reach us directly from that point. That is to say, photons from a redshift of 1000 are the oldest that we can see directly. The cosmic background radiation is made up of photons from that ancient light cooled now to the microwave part of the spectrum.

The universe much beyond a redshift of 1000 is unobservable. We cannot see through the fog, as there is a photon barrier or last scattering surface at that redshift. The microwave photons that comprise the cosmic background radiation come from that last scattering surface. Very small variations in the pattern of microwave background radiation reveal to us the density perturbations of the early universe from which structure grew and evolved. We are looking at the universe at redshift 1000 when we look at the background radiation and the fluctuations that we see reflect the low level to which they had grown—about one part in a hundred thousand variance from place to place—at that time.

This description of *what* happened is physically sound and fits the description of the facts, but the good journalists' questions of *how* and *why* it happened as it did brought only embarrassed silence, because the conditions needed seemed to require such incredible fine-tuning. The postulated big bang had to have been "just right"—just right with regard to the ratio of photons to hydrogen atoms, just right with regard to the level of the fluctuations, and just right with regard to its overall density. In outlining these problems and the ingenious solutions that have been proposed, we look to Jim Peebles' book *Principles of Physical Cosmology* (Princeton, 1993) as our guide. The solution to the fine-tuning problem

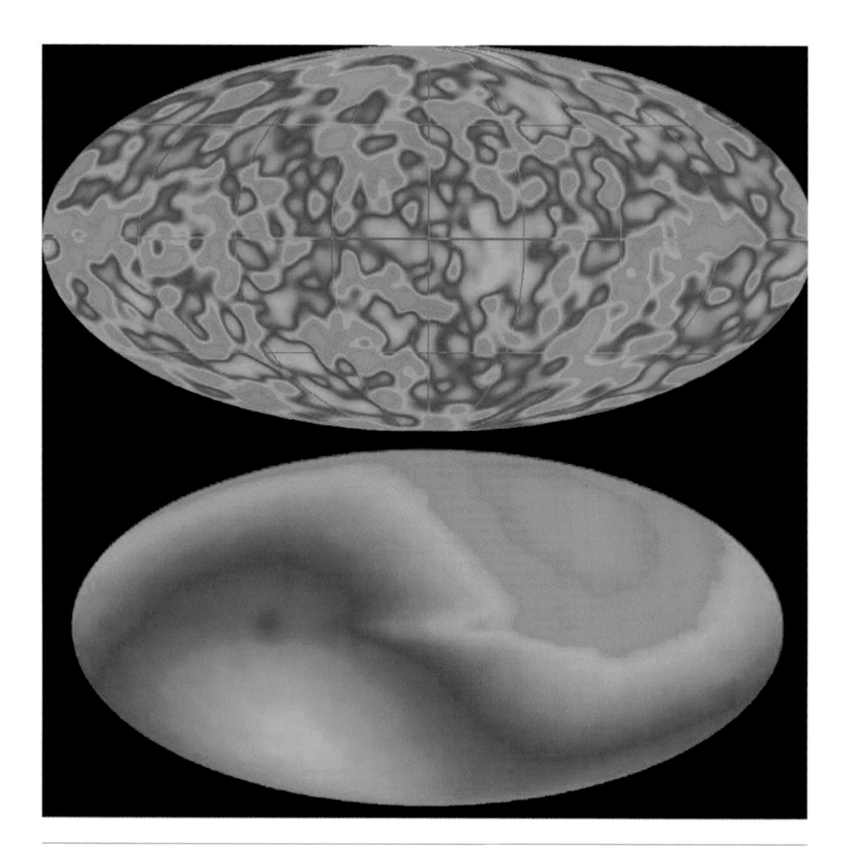

Figure 5.5. The full sky map (*top*) showing for the first time (1989) the fluctuations in the cosmic background radiation field that seeded the growth of structure in the universe. Observations from the COBE (COsmic Background Explorer) space telescope. The lower panel shows the dipole variation over the sky (removed from top panel) caused by our motion through the cosmos. (NASA/COBE)

was called "inflation," the idea that the universe had an early period of exponential growth in its earliest phases that *preceded* the conventional Big Bang. While we may only be exchanging one mystery for another, the general belief among cosmologists is that the theory of cosmic inflation does make it easier to understand the nature of the Big Bang. And, as we have seen, there are two separate questions: why is the observable universe so uniform on large scales (as we shall soon see, this simple fact becomes quite puzzling when one thinks about it), and how did the evident non-uniformities arise? It turned out that these two seemingly opposite questions were both to be addressable by the same process.

■ Cosmic Inflation

Even apart from questions about the growth of perturbations, the successful background model for the evolution of the smooth component presented paradoxes. By the end of the 1970s, a so-called standard model of the hot big bang had been formulated, and it seemed to work. But it posed deep puzzles, new puzzles as profound as those that it had solved.

The first puzzle was the horizon problem. In a hot big bang containing only matter and radiation, the universe flies apart before it can come into equilibrium. Putting that another way, in the simplest picture that one can imagine, all parts of the universe were in contact with all other parts. But, in the standard big bang model of Lemaître and Hubble, there are regions of the universe that are widely separated today, which have never been in causal contact in the past. When we look to the most distant quasars in the northern sky and the most distant quasars in the south, we are looking at objects that cannot have seen one another. The distance between them is greater than the distance that light can travel in the time from the big bang until now.

So the horizon problem is this: why is the universe uniform and isotropic now, when there appears to have been no means in the past to communicate such uniformity? When we look left or right, up or down with our biggest telescopes, we see parts of the universe that have not seen each other. If you pour water into a glass, it does settle to give a horizontal even layer. But, instantly, right after it has been poured, when water on the one side does not even know how high the other side is, how could one get a level surface? It could not and does not happen that the water level becomes instantly flat. It takes time and communication (by pressure-driven sound waves within the glass) for the water to settle and the left and right sides to reach the same level. Both sides of the glass feel the same gravity, it is true, but initially water on one side does not know how high up in the glass is the water on the other side. In a high-speed movie we would see the water slosh back and forth for a while as the pressure from one side was communicated to the other.

How, then, in the universe can very distant parts sense how to be similar to one another? Enough time has not elapsed for the information about one side to have reached the other. It is a bit like the classic paradox of two male identical twins, separated at birth, having grown up without communication at opposite points of the globe, who are brought together at a halfway point. In the fanciful and oft repeated account, they meet at a restaurant for dinner and find that they are dressed in identical suits, shirts, and ties! Well, repeated observations had shown that our real universe has this bizarre property—the northern and southern quasars, while individually different, are statistically identical. It took until the mid-1970s for astronomers to realize the strangeness of the situation, and this was only one of the underlying curiosities.

Second, we have the flatness problem, which arises because the present mean density of the universe is incredibly close to the critical density at which the geometry of the universe is termed "flat." We learned about the critical density in chapter 3: it is the density that the universe would have (neither more nor less) so that its velocity of expansion could *just* be enough to overwhelm the gravitational force pulling it back—so the expansion could last forever. This is another example of astonishingly acute fine-tuning. If the universe had been on average 0.01 percent more dense than it was at a redshift of $z = 10^6$, then it would already have collapsed to an immensely dense state. But, if it had been 0.01 percent less dense, then it would have flown apart to a tiny present density that would not have allowed galaxies, stars, and planets to form.

Some cosmologists have invoked the anthropic principle as reasoning to explain away the problem: the universe has the properties it has because we live in it and can observe it. A substantially different universe would not have allowed the evolution of sentient beings. True enough. However, the ancient philosophers might have regarded this argument as illustrating the classic error of begging the question. The argument is in any case unscientific, as it is not capable of being refuted by any observation or experiment. The anthropic argument has also been used to address the quite distinct question of fine-tuning noted earlier with respect to the degree of irregularity in the universe—the Goldilocks problem. Why were they just right and neither too large nor too small? While many esteemed scientists have other views on the matter, it has seemed to the authors of this volume that, while the anthropic principle is interesting as philosophy, the fact that it cannot be falsified puts it outside of the realm of science.

A third conundrum concerns the absence of exotic particles in our universe. On a straightforward view, the early universe should have produced a copious abundance of magnetic monopoles: stable, heavy particles that can be imagined as dense knots of magnetic field. This is clearly not the case.

Alan Guth was one of the first to propose an astounding solution to this triad of deep and tricky questions back in 1980. At the time he was a young physicist at Stanford University, searching for a permanent post. Guth had a string of degrees from MIT, and then progressed through Princeton, Columbia, Cornell, and Stanford on a succession of short-term appointments in particle physics. At Columbia he had focused on magnetic monopoles, and he became something of an expert on these (nonexistent) entities. Two lucky breaks then changed the course of his life. On the hilly Ivy League campus of Cornell University in 1978, he listened with great interest to a lecture on the flatness problem given by a visitor from Princeton, Robert Dicke.

Dicke intrigued young Guth with his assertion that something important must be missing from the standard big bang theory. In a complete theory there must exist some physical process that would make reasonable the apparent fine-tuning of the universe. Guth's next step resulted from his attendance at two lectures by Steven Weinberg on grand unified theory, a concept that three of the prime forces of physics (electromagnetism, the strong force, and the weak force) can be combined into a single theory. From Weinberg, Guth learned about phase transitions in the early universe.

In what has later been described as a flash of insight, Guth realized that a number of cosmological puzzles might be solved with a single concept, namely that the early universe underwent a short-lived period of inflation, during which the expansion ballooned at an extraordinarily rapid, exponential rate.

We start our explanation with the concepts of latent heat and phase transitions (transitions of water to ice or water to steam are domestic examples of phase transitions). Guth's own path to the inflationary universe had involved the notion that the universe underwent super-cooling through a phase transition. An easy route into this complex theory can be found by imagining a fish pond covered with ice. Even if the air temperature is well below freezing, the temperature of the liquid water below the ice remains at freezing point, until the pond is frozen solid. That is because the formation of ice, which is a phase transition from liquid to solid, releases latent heat, which stabilizes the nearby water, keeping it exactly at the temperature of the transition from water to ice. Guth realized that, if the very early universe had gone through such a phase transition, then a tremendous amount of energy would have been released. Obviously the transition would need to be far more exotic than water or some other liquid turning into a solid. Guth, a rookie cosmologist, started with the universe in a symmetrical state, having nuclear and electromagnetic forces in harmony. This balanced universe contained a great deal of latent energy. But, as the universe expanded, its temperature fell until it reached a critical value, at which point the symmetry became broken. Matter assumed a new state in which the forces binding the nucleus were overwhelmingly larger than the electromagnetic force, just as pertains today. And the latent heat that was released drove the universe through a period of stupendous expansion.

Guth speculated that this transition began only 10^{-34} seconds after the onset of the big bang. At that stage the characteristic scale of length, the distance that light could travel in 10^{-34} seconds, was only 3×10^{-26} meters. In the exponential expansion, which lasted only from 10^{-34} to 10^{-32} seconds, Guth showed that the length scale expanded by a factor of 10^{43} (for the technically minded, $\sim e^{100}$), to 3×10^{17} meters. At our current epoch this length scale of 3×10^{17} meters (back then) has now grown, in the normal expansion of the universe, to 3×10^{42} meters, which far exceeds the horizon of our observable universe, which is $\sim 10^{26}$ m. During this period (10^{-34} to 10^{-32} seconds), as long as the phase change was going on, it seems that space itself inflated. We can pursue our homely analogy by considering what happens to a glass bottle, filled with water, that is placed in a freezer. As we all know, ice is less dense (takes up more space) than the equivalent amount of water. So a time lapse photograph would show the glass bottle suddenly shattering when the water crystalized into ice at the relatively benign phase change of water into ice.

The implication of cosmic inflation is that our observable universe arose in a very tiny region, where all the parts were close enough together to be in causal contact. Or, to return to our other metaphor, we learn that the twins separated at birth carried with them instructions from their joint mother on dress code that they followed to the letter. In fact, parts of the universe that are now far apart were close together in the pre-inflation epoch. That solves the horizon problem: our universe was isotropic before inflation and that is why it appears isotropic on large scales now (isotropic means that its properties appear to have the same value in all directions). The awesome exponential expansion explains the flatness problem. The huge amount of stretching of the geometry straightened out any kinks and bumps that may have existed initially. With its temperature changed dramatically, and its expansion accelerated, the universe no longer would produce a vast density of magnetic monopoles.

Thanks to his strong background in particle physics, Guth addressed problems that the cosmologists had not come to grips with. The inflationary universe is predicted to contain Higgs bosons, postulated elementary particles that the Large Hadron Collider (LHC) in Geneva is designed to detect. The LHC is the biggest and most expensive physics experiment ever attempted, and it will fleetingly replicate the enormous temperatures that prevailed soon after the Big Bang. The recent, 2012, apparent detection of the Higgs boson has provided real support for this picture of what happened in the early universe. But, even though the postulated particle may have been found, there remain problems with this solution to the paradoxes implicit in the standard big bang picture.

Not surprisingly, Guth's revolutionary paper (published 1981) was found to contain flaws, but other physicists quickly resolved many of these. A consensus soon arose that inflation did solve many more problems than it creates. However, it turns out that even now, three decades later, no single version of the inflationary paradigm has been developed that is free of inconsistencies and additional paradoxes. It appears that the inflationary paradigm is still, as cosmologist Paul Steinhardt (a co-inventor) has remarked, a work in progress. The more deeply individual variants of this attractive picture are studied, the more problems are uncovered. Thus, while we are confident that the current inflationary picture contains some fundamental truths, we know that it is far from the final answer to the puzzles of flatness and isotropy. Undeterred, let us now examine how inflation has an impact on the problem of the origin of structure.

The universe before inflation was a quantum world. How long did the initial instant last? Less than 10^{-43} seconds. How do we know that? Well, Einstein's general theory describes the history of the universe right back to 10^{-43} seconds, at which time the universe is so dense that we have to use a *quantum* theory of gravity instead to understand it. Since the 1930s, it has been known that Einstein's general relativity fails at this point. But there is no such quantum theory of gravity, only a lot of extremely intelligent speculation, by the string theorists. Maybe the LHC will help here, if its users discover a plethora of new particles. But we know with certainty that general relativity runs into the quantum barrier at this early epoch (10^{-43} seconds) and that Einstein's theory cannot handle it.

One hope the theorists have had in the inflation scenario is the possibility of finding the origin for the perturbations from which cosmic structure later emerged. Let's go back to our fish pond analogy for a moment. When the freezing starts, it does so through a process of nucleation: the ice forms in separate discrete solid particles. Could the phase change in the early universe have produced something lumpy? In the original version of the inflation theory, the primeval chaos curdled and produced huge perturbations, much larger than those found in the real universe. However, a revised version of the theory addressed these problems. While the details remain obscure, there is a general belief that the large-scale cosmic fluctuations, which provide the structure of our current universe, arose from tiny quantum fluctuations in the early, pre-inflation universe. Cosmic inflation then stretched them by the giant factor of $\sim 10^{43}$ to the present cosmic scales.

Spelling out this process also tells us the amount of fluctuation as a function of the scale. Smaller length scales fluctuate with large-amplitude waves, but the fluctuations from one large piece of the universe to another are much smaller, since the small-scale up and down density fluctuations tend to cancel out when we change our focus to the larger picture. The neat solution, to which Steinhardt contributed, not only gives the scale of the fluctuations, but also predicts (approximately) the detailed spectrum of perturbations, the relative amplitudes of longer and shorter waves to be seen in the cosmic background radiation. In fact, a decade earlier, in the early 1970s, three physicists had separately arrived at the same hypothesis. Edward Harrison, Y. B. Zeldovich, and Jim Peebles (HZP) had independently proposed that the spectrum of perturbations must be "scale-free." What this meant was that at all times the perturbations on the largest scale that matters, the scale of the horizon of our observable universe (with radius equal to the speed of light multiplied by the elapsed time since the Big Bang), would have the same (small) amplitude. If this were not the case, then, either in the past or in the future, cosmic perturbations would be so large that larges pieces of the universe would collapse to black holes. The scale-free spectrum, which avoided this unpleasantness, was also the spectrum mandated by the inflationary scenario.

When data from the Cosmic Background Explorer (COBE) satellite launched in 1989 was analyzed (see fig. 5.5), the prediction immediately seemed to be confirmed. Subsequent information from later satellites, showing the same sky map but with far more detail, has steadily improved the correspondence between the predictions of inflation, HPZ, and the real world. But again, we are jumping too far ahead.

Our colleague Malcolm Longair (Cambridge) has described the quest for the origin of primordial fluctuations as "one of the major growth areas of theoretical cosmology." Another of our Cambridge colleagues, Stephen Hawking, has worked in the field since its inception. In 2000, Andrew Liddle (University of Sussex) and David Lyth (Lancaster University) gave this neat appraisal:

> Although introduced to solve problems associated with the initial conditions needed for the Big Bang cosmology, inflation's lasting prominence is owed to a property discovered soon after its introduction. It provides a possible explanation of the initial inhomogeneities in our Universe that are believed to have led to all the structures we see, from the earliest objects formed to the clustering of galaxies, [and] to the observed regularities in the microwave background.

■ The Seeds of Cosmic Structure Are Discovered

With this introduction, let us now turn to how the long sought seed fluctuations were found. In chapter 4 we introduced the concept of a hot big bang universe and showed how the discovery of the leftover radiation from this cataclysmic event established the framework for our basic cosmology. Now let us bring this part of the story up to date to show how the COBE satellite was launched and what it achieved. Although, as noted, the actual discovery of the cosmic microwave background (CMB) occurred in the mid-1960s, a handful of cosmologists had given passing reference in the 1950s to the background temperature of the universe. Furthermore, radio astronomers had made the first unsuccessful attempts to see if a universal background could be detected. However, the literature is either silent or understated on the cosmological implications of such a background. Professional astronomers displayed almost no interest, except for George Gamow who, as we have already noted, did take the cosmological applications seriously.

In the summer of 1953, Gamow lectured on cosmology at a symposium hosted at the University of Michigan, Ann Arbor. In the published account of that symposium, Gamow gave his version of the formation of the elements in the hot big bang. Early in the lectures he floated the idea of a sea of thermal radiation that cools as the universe expands, and he considers what effect the radiation might have on the expansion of the universe. This is the first mention (known to us) in the literature of the cosmic background radiation field. Gamow touched on the issue of galaxy formation, and the large-scale structure evidenced by rich clusters of galaxies. Significantly, he suggested that the mass density of the universe at the transition when the universe became matter-dominated would have set the mass at which galaxies would form. He arrived at that conclusion by noting that clouds condensing under gravity in the early universe would have a minimum size set by the Jeans mass and length. The minima turned out to be roughly the mass and size of normal galaxies.

Thus, Gamow was the pioneer who foresaw that the background radiation would have implications for the formation of structure. Unfortunately he did not publish this in the *Astrophysical Journal*, so no notice was paid to his thoughts. In the period 1948–65, the debate in physical cosmology was dominated by the two rival models: the big bang and the steady state, while observational cosmology remained a search for two numbers.

In 1964, Penzias and Wilson changed everything with the chance discovery of the background radiation (fig 4.6). It was clear that Hoyle's steady state universe could not be our universe, and that the cosmic background radiation was a fossil from the Big Bang. Theorists were not slow to realize that the apparent structure of the radiation on the sky would provide a glimpse of the early universe, freely propagating to us from an era when the universe was structurally simpler than today. Higher and lower density patches at the early epoch would have been hotter or colder, the radiation from them brighter or fainter, and in color bluer or redder. Looking back to this epoch we would see the fluctuations, anisotropies—the peaks and the troughs—in the background radiation, and these would provide a unique measure of the primordial density fluctuations from which the present large-scale structure arose. The theorists noted that the inflation era provided a means by which the infinitesimal quantum fluctuations of the earliest universe could be amplified to cosmic scales by means of exponential inflation. However, in the two decades following discovery, the microwave radio telescopes had not achieved the requisite sensitivity or angular resolution to see the expected fluctuations. The path to obtaining high-resolution maps of the background radiation turned out to be tortuous. Experiment after experiment, using ground-based radio telescopes, failed to find the fluctuations that the theorists had said *must* be there.

But, in 1974, NASA transformed the situation with an announcement of opportunity for astronomers to fly missions on a small *Explorer* class satellite. NASA had been persuaded that only by going into the calm and noise-free environment of space would we be able to survey the background radiation with sufficient steady precision to see the promised "face of God." NASA received more than one hundred proposals for *Explorer* missions, three of which concerned the background radiation. The contest was won by the proponents of a different mission, the Infrared Astronomy Satellite (IRAS), which was ultimately a huge success. However, NASA did not discard the proposals for studying the microwave background. Far from it. In 1976, they invited the bidders to form a consortium to define a mission. NASA suggested using a satellite in polar orbit launched from the Space Shuttle. Its name was the Cosmic Background Explorer, or COBE for short (see fig. 5.5). NASA approved the mission, but cost overruns on IRAS delayed the start of the construction of COBE until 1981. Launch was slated for 1988, but the *Challenger* tragedy of January 28, 1986, led to the Shuttle being grounded. COBE eventually soared aloft on November 18, 1989 on a Delta rocket. The engineers and physicists who had designed COBE were understandably nervous at the launch: they asked themselves, would a spacecraft that had spent four years in storage work?

Yes, actually it did.

In January 1990, one of us (Mitton) attended the 175th Meeting of the American Astronomical Society, held in Crystal City, Virginia. At 2 p.m. on January 13, John Mather of the NASA Goddard Spaceflight Center stood at the podium to deliver his ten-minute talk. He spoke calmly about the COBE satellite, explaining to an audience of almost 1,000 people that the mission was going smoothly, and saying that it would take a year or more for the COBE instruments to map the entire sky. But he already had an important result, a spectrum (showing the intensity of the radiation versus the wavelength) of the radiation based on observing a small patch of sky for just nine minutes. To conclude the lecture, he picked up a transparency and placed it on an overhead projector so that its image appeared on a giant screen. “Here is our spectrum. And this is a black body curve that connects all of the data points.” For a moment a hush descended, then there were murmurings. Next, people began to applaud. Then everyone stood up, clapping wildly and enthusiastically. Said Chuck Bennett, a member of the COBE Science team, “I’ve never seen anything like it at a scientific meeting.” The COBE data were a perfect fit to the theoretical curve for radiation within a “black body” as derived by Max Planck in 1901! The temperature of the radiation was 2.736 ± 0.017 K. This discovery cemented the Friedman-Lemaître-Gamow picture of the evolving, hot big bang universe.

The same day, a few hundred miles north of Washington, David T. Wilkinson, a Princeton professor (and younger colleague of Bob Dicke), who was a leading member of the COBE team, was presenting the same material to an audience which included the other author of this volume (Ostriker) at a special Princeton colloquium. Again there was wild, spontaneous applause when the radiation spectrum was shown. In this venue, a map of the fluctuations in the background was also shown and a debate broke out instantly as to whether or not these were large enough to account for the observed structure in our local, low redshift universe. Yes was the answer. But one contributor from the floor interjected a critical qualification. The observed perturbations were big enough, if—and only if—there was a great deal of "dark matter" in the universe, much more dark matter than ordinary matter, in order to gravitationally boost the rate of growth to the needed level. It was an exciting moment in the history of science!

The most detailed maps followed in 1992, about two years later. They showed a "dipole" variation over the sky, having a variation level of about 0.1 percent. One octant of the sky was about 0.1 percent brighter than the average, and the opposite octant was about 0.1 percent fainter than the average (see fig. 5.5 *bottom*). This is caused by the motion of the Milky Way galaxy relative to the rest frame of the universe, as defined by the radiation. It was easy to subtract off this component, in order to reveal the large-scale irregularities known as anisotropies. These are of the order of a few parts per 100,000, and they preserve information on the formation of structure in the evolving universe. The 1992 COBE announcement of the background radiation anisotropies generated enormous publicity in the media across the globe, and it seemed to provide final confirmation of the lumpiness predicted by theory and required for galaxies to form.

Also, the observed dipole variation showed directly the motion of our frame of reference with regard to the universe as a whole. The dipole variation shows that the Local Group of galaxies is travelling toward the Virgo cluster of galaxies at about 600 kilometers per second. That motion (fig 5.5 *bottom*) is caused by the gravitational attraction of the Virgo cluster.

The COBE satellite established three fundamental facts about the universe. It found the predicted radiation left over from the hot big bang; it detected the perturbations needed for the growth of structure, confirming the expected spectrum of perturbations; and it found the (non-moving) rest frame of the cosmos. Rather good work.

COBE made a huge impact in the scientific community, so much so that the competition for an even better instrument began almost immediately and was won by the Wilkinson Microwave

Anisotropy Probe, or WMAP, launched on June 30, 2001. This mission was named in honor of its leader, Dave Wilkinson, the core member of the COBE team mentioned earlier, who died in September 2002 before data from the second mission could be analyzed. The WMAP scientific team was led by two other Princeton astronomers, Lyman Page and David Spergel, who had been associates of Wilkinson. To address key questions in cosmology, WMAP re-measured the small variations in the temperature of the cosmic microwave background radiation first detected by WMAP, and it did so at higher accuracy and higher spatial resolution. The variations are miniscule: one part of the sky has a temperature of 2.7251 K, while another part of the sky has a temperature of 2.7249 K. Higher temperature corresponds to higher density, so the hot spots found and seen in figure 5.6 would be sites for the later formation of galaxies and clusters of galaxies. These exquisite measurements revealed the size, matter content, age, geometry, and fate of the universe. In the next two chapters we will turn to WMAP's contributions to precision cosmology. Here we note that WMAP revealed the primordial structure that grew to form galax-

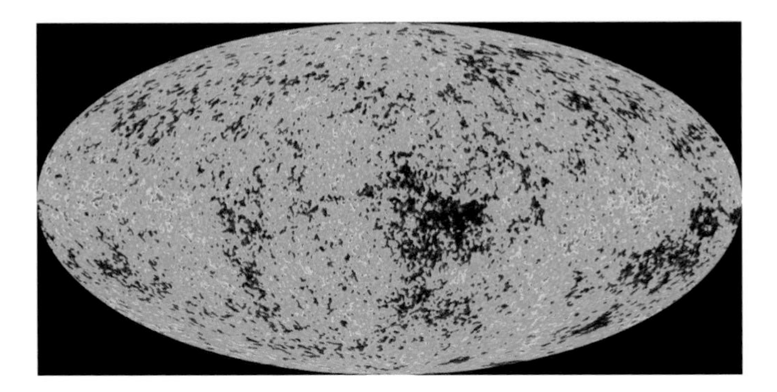

Figure 5.6. Fluctuations in the cosmic microwave background observed by the Wilkinson Microwave Anisotropy Probe. The differences between bright and dark regions are greatly amplified (by a factor of roughly 100,000 to make them apparent in this image and in (5.5 top) as well. Bright spots will become, over cosmic time, clusters of galaxies and dark spots, voids in the distribution of galaxies. (NASA/WMAP)

ies. That discovery challenged scientists to derive clean theories for the origins of galaxies and the larger cosmic structures. Let us see if we can explain *how* gravity acted to make these initially tiny fluctuations grow into the structures that we see around us.

■ Closing the Loop: How Do Seeds Grow to Galaxies?

Galaxies have a mean density of roughly 0.02 solar masses per cubic parsec, or about one hydrogen atom per cubic centimeter. This is about four million times more than the mean density of ordinary matter in the universe and, as we have noted earlier, is closer to the mean density in the universe long ago, when the galaxies were formed. Stars within galaxies have a density twenty-three orders of magnitude greater than galaxies. This is an almost incredible degree of concentration. So gravitational instabilities have their work cut out for them to be able to produce this level of inhomogeneity from such small beginnings. And the small beginnings absolutely have to be in place at the outset, because the universe is already too large for the *initiation* of structure formation once the inflationary era is over. It might be imagined that, during the eons from the inflationary era to the redshift of $z = 1000$, there would have been considerable growth. However, detailed calculations show that not to have been the case; the tight coupling between ordinary matter and the very hard to compress background radiation field holds the matter in place, not letting it move very much in responding to gravitational irregularities. We can begin therefore with the maps released by the WMAP science team in 2010.

Now, for the mathematically inclined, we can address the physics of how these tiny seed fluctuations of a few parts per 100,000 variation from place to place could grow, after a long period of stagnation, to the current inhomogeneous state seen in the large-scale structure maps of the universe. In the 1960s, the Moscow and Princeton groups independently solved this problem of how the fluctuations would develop and grow over time, and the mathematics needed to treat fluctuations over the whole sky is quite complex. However, if one wants to look at a tiny representative piece of the universe, there are elementary methods available as we saw in chapter 3 and in the appendix. The easiest way to make the calculation is to go back to the simple problem treated in the appendix of a small expanding sphere and imagine a little region that was slightly denser than the surrounding medium at an early time and to see how that region evolves.

What happens in this case was solved in a neat short paper by James Gunn and J. R. Gott in 1972. They showed that, even if the universe has a density lower than the critical density (as our universe almost certainly does), then the little sphere, which we studied earlier, will ultimately collapse if and only if it has a density within itself that is ever so slightly greater than the "critical density"—the magical, mean density for a universe which would just barely have the energy to expand forever and not to recollapse. Intuitively, this is reasonable, even obvious. As we noted earlier, Newton had proven that a medium of uniform density outside of a spherical region produces no forces on the sphere. So the little sphere simply acts as its own universe with more than the critical density all by itself; it expands, reaches maximum radius, stops, and then collapses. Let the mass be a tiny fraction f greater than the critical mass; then equation (A12) can be rewritten as $v^2 = 2(GM/r_0)[(r_0/r) - f]$.

This simple equation tells all. Here M is the mass of our little sphere, v is its velocity of expansion, and r_0 is the radius of the little test sphere when we first look at it (at time $t = t_0$), while it is expanding rapidly with the rest of the universe. But as time proceeds and r increases, the right-hand side of the equation drops steadily and the rate of expansion slows. Our sphere finally stops expanding when the right-hand side of the equation is reduced to zero. We see that the radius has grown by the big factor $1/f$ when the expansion stops. Let us call that time t_{max}, when r reaches $r_{max} = r_0/f$.

What follows from this? Well, the sphere cannot just stay stationary, since there is nothing to counteract gravity, and it will

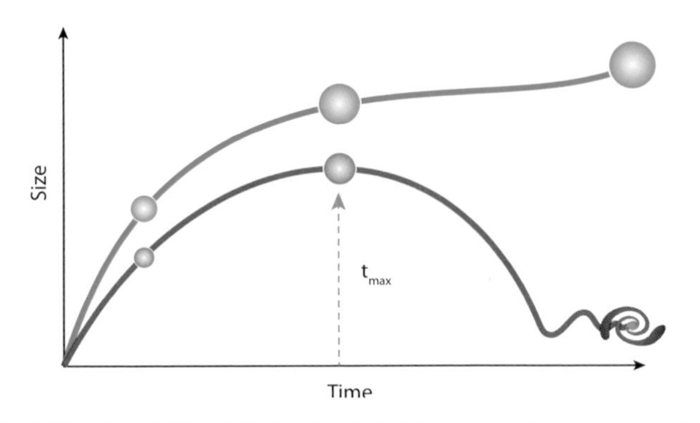

Figure 5.7. While a representative spherical piece of the average universe grows with time in the fashion of the green sphere, the red piece, with density greater than the critical density of matter, expands more slowly, reaches a maximum size at time t_{max} and then collapses to a self-gravitating galaxy at time $t = 2\ t_{max}$.

start to contract again. The motion is completely reversible in time. The little sphere is acting in precisely the same way as a sphere in a "closed" universe, so it can and will collapse back to a very small volume by the time $t \to 2\ t_{max}$. The density within it could easily be a million times more than the density in the rest of the expanding universe by this time. Now we are looking at a collapsing sphere of very high density and can use the Jeans analysis, with which we first began.

Needless to say, the Jeans analysis shows that this dense gaseous sphere could in fact fragment into stars, but describing that process is turning to yet another complicated story and one still imperfectly understood, star formation. However, suffice it to say, we know that dense self-gravitating clouds of gas are in fact turned into stars by processes that we still do not fully understand, even though we can see it happening before our eyes in, for example, the Orion Nebula. So we may assume that the collapsing cloud will form into an assemblage of stars that somehow re-arranges itself into a galaxy!

This description in words does not do justice to the beauty of the process, but movies made from the numerical simulations of the process are quite gorgeous and confirm the story presented here. Figure 5.8 shows twelve snapshots from a movie made from a numerical simulation of galaxy formation. The principal difference between the complex reality shown in our snapshots and our cartoon little sphere is that there is not one little over-dense sphere in the real case but many smaller interacting and merging little dense regions. However, overall, the simple picture is correct.

Well, what happens next? One galaxy is made, but so are many others and, if a galaxy finds itself in a region with an above average density of galaxies, a density so high that it is again above the critical magic value, that region itself, containing perhaps a few galaxies or perhaps a thousand, will itself start to contract and finally collapse into a bound group. Our Local Group, which contains the galaxy, the Andromeda nebula, the Magellanic clouds, and a several other systems, is a small version of this. The nearby Virgo cluster of galaxies, in whose outskirts we live, is a bigger example, and the great Coma cluster of galaxies (fig. 6.1) is a grand example.

When computations are made of what happens to many particles interacting via Newton's laws in an expanding universe, the gravitational instabilities just described happen in just the way described. But, and this is a very big but, it does not work at all if we only put into the computation only the matter that we see in the stars and gas that is easily measurable. If we have only the visible stars and the detectable gas, then the f in the last equation is *negative* and the expansion never reverses course. To make the little region have a density larger than the magic, critical value, we *require* the extra dark matter. The interjection from the floor at the scientific meeting at which the COBE results was presented was correct! It only works out right if we add extra matter to make the gravitational forces large enough to drive these instabilities. What is this extra stuff? Dark matter. And a nice point of consistency is that we need for this purpose just the amount of dark matter that had been discovered by other means.

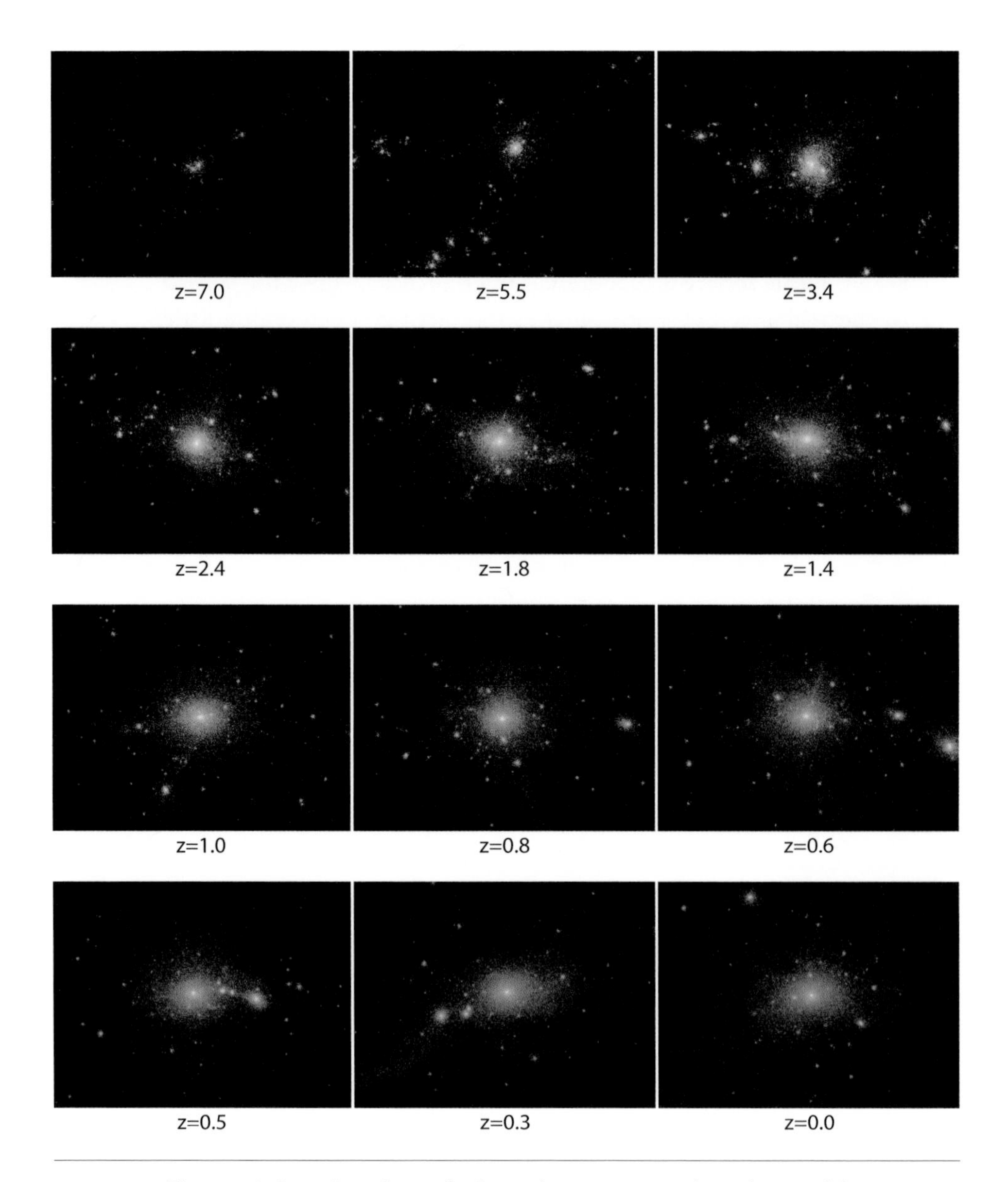

Figure 5.8. Snapshots from a high-resolution numerical simulation of the growth of a giant elliptical galaxy from cosmological initial conditions by Thorsten Naab (2007). Young stars are bluer and old stars are more red. The simulation starts at redshift $z = 7$ and ends at the present time. (Thorsten Naab, Max-Planck Institute for Astrophysics, Munich)

We have come a long way. The hot big bang had been postulated to provide the background for the formation of the light elements and the expanding universe of galaxies, and then the cosmic background radiation was fortuitously discovered. Next, fluctuations in the background were postulated to provide the seeds for the growth of galaxies, and, after great efforts, these fluctuations too were discovered. Now we come to the physical question of how and why this all happened and to the most recent and dramatic discoveries. The primary components of the universe, and the ones needed to explain what we have already discovered, are essences: *dark matter* and *dark energy*, not found (yet) on the Earth but dominating the heavens. Again we have gone too far. We end this chapter with the growth of structure comprehensible and dark matter foretold.

Chapter Six

Dark Matter—or Fritz Zwicky's Greatest Invention

■ How the Earth Was Weighed

What is dark matter? Do we really need to believe in its existence, or is it simply the invention of excessively ingenious minds? We have had a rather casual introduction in the last chapter to this principal component of the universe, as the stuff that makes gravity act properly, to force the growth of all the structure that we observe around us. But that is not at all the way that dark matter was found. What we shall show here is that there was not one argument for this strange material, nor two, but many distinct observational clues, and they all pointed to the existence of material with the same properties: absorbing and emitting no light, but exerting and responding to the force of gravity, and, in amount, roughly six times as much as the ordinary chemical elements from which we, the Earth, and the stars, are made.

The observers' initial discoveries of dark matter were completely independent of the dynamical investigations presented in the last chapter. Their efforts to understand the growth of structure in the universe came decades later than the seminal 1937 observations that we will describe here. The tale is not the linear one sometimes presented in the history of science, where greater and greater levels of information lead to a gradual consensus. Rather, there were brilliant pre-discoveries that were ignored and a variety of different lines of evidence, frequently hotly disputed and often initially misinterpreted, that after much debate finally produced a tipping point when the unconventional became the conventional, authorized belief.

Chronologically, we start with Fritz Zwicky, already introduced to you as the brilliant but zany astronomer, born in 1898 in Varna, Bulgaria, the son of the Swiss ambassador to that country. Zwicky, one of the most original astrophysicists of the last century, was educated at the Swiss Federal Institute of Technology, and moved to Caltech in 1925, where he worked most of his life at the Mount Wilson and Palomar Observatories. He was truly the discoverer of dark matter, even calling it *dunkle Materie*, in his paper of 1937, in which he analyzed the dynamics of clusters of galaxies. However, it was decades before his great discovery was fully understood. Basically, he found, labeled, and cataloged giant clusters of galaxies and tried to understand what glue could hold the galaxies in them together. The galaxies in these clusters had enormous velocities, typically 1000 km per second. Yet they had not flown apart and dispersed. Some force was keeping the clusters intact. The obvious guess as to what that force must be was the gravity of our old friend, Isaac Newton: the same force that holds the Moon in orbit around the Earth and the planets in orbit around the Sun.

But, in order for gravity to be strong enough to hold together these giant clusters (see fig. 6.1), there had to be much more mass present than was apparent from the number of galaxies seen, allowing for each galaxy the normally assumed mass. The force of gravity is proportional to the amount of mass that is present, and, if we were to assume that the average star seen in these systems had the mass of our sun, then the total gravitational force from them would be deficient by roughly a factor of 100 to do the job of binding the clusters of galaxies. So Fritz Zwicky, never modest in his speculations, postulated that there was something else in the clusters, namely dark matter, that provided the extra mass and was responsible for stopping the galaxies in the clusters from flying apart.

Figure 6.1. Central region of the nearest great cluster of galaxies, the Coma Cluster. This is the prototype of the systems that Fritz Zwicky found in which the galaxies had such high random velocities (~ 1000km/s) that extra strong gravity—possibly due to dark matter—was needed to keep them from flying apart. (NASA)

Before describing Zwicky's extraordinary proposal, it will be helpful for us to review the fundamental question of how astronomers find the masses of planets, stars, and galaxies. Zwicky's method for estimating the amount of dark matter in galaxy clusters is not in principle any different from how the orbit of the Moon was used to determine the mass of our Earth in the eighteenth century. Once again, we will use high school mathematics to show the way, as the analysis does not require more than that.

Let us start by calculating the mass of the Earth. The easiest way, and one available to the ancients, would be to take its known radius, 6400 km, and use that to calculate the volume. The result is 1.1×10^{21} cubic meters. Now multiply the volume by the mean density of the Earth to find its mass. What should we use for the density? Again, imagine that we are making this calculation at the time of Hipparchus: we will take, as a first estimate, the density of rocks that seem to come out of the interior of the Earth—granite or basalt. These primitive rocks have a density of about three times that of water, so our estimate for the Earth's mass would be roughly 3.3×10^{24} kg. The truth is about twice this number (6.0×10^{24} kg). Our first attempt at estimating the mass was a good one, but it is a bit low because of the iron and other high-density materials in the center of the Earth.

But how do we know what the true mass of the Earth is? To find it we use the force of gravity and Newton's law of universal gravitation: all matter exerts a force on other masses proportional to the product of the two masses and inversely proportional to the square of the distance between them. So the general form of the Newtonian force law, which we have used before, is $F = GmM/R^2$, where G denotes Newton's universal constant of gravitation, M and m are the masses, and R is the distance between them. Let's now do a thought experiment.

First consider an idealized experiment, shown in figure 6.2, where two really massive balls, mass M, are suspended from the ceiling of a laboratory with separation D. Let's think about the forces acting on each ball. There are two forces to consider. In the vertical direction each ball is being pulled down vertically by the force F_v of Earth's gravity, and that is $F_v = GMM_e/R_e^2$ where M_e is the mass of the Earth and R_e its radius. In the horizontal direction each ball is pulled horizontally toward it neighbor with a force $F_h = GM^2/D^2$. The two lead balls do not hang down in an absolutely vertical direction. That's because they will incline slightly toward each other: they are mutually attracted. The small angle that the cable to each ball makes to the vertical depends on a ratio: the mass of the ball divided by the mass of the Earth, and, the square of the distance between the balls divided by the radius of the Earth. That ratio is $r = (F_h/F_v) = (M/M_e)(R_e/D)^2$ and it is equal to the small angle each ball makes to the vertical. Let's assume the angle can be measured, in which case everything is known except the mass of the Earth. So in this thought experiment we have to solve an easy equation to weigh the Earth.

The beauty of this demonstration is that we do not need to know the value of Newton's constant, G. That's because the calculation uses ratios, and the gravitational constant drops out. Neat! But then, having found the mass of the Earth, we can find G as well by

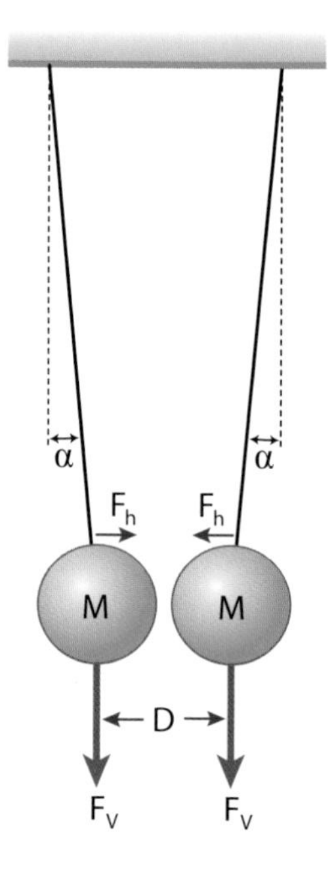

Figure 6.2. Two lead balls are influenced by the downward gravitational attraction of the Earth and in the horizontal plane by their mutual attraction. The small angle of inclination to the vertical of the suspension cables is the ratio of the horizontal and vertical forces, from which the mass of the Earth can be calculated.

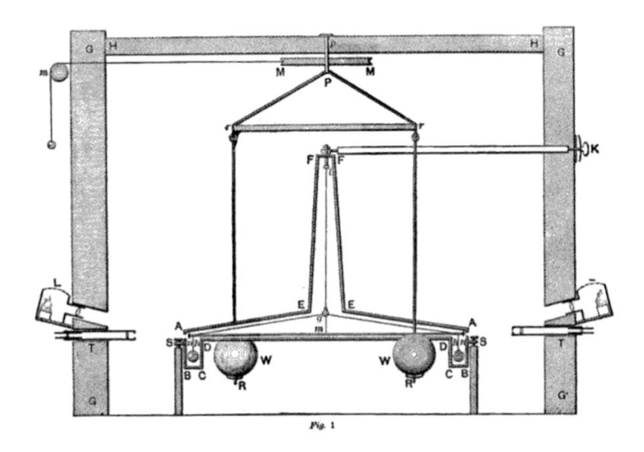

Figure 6.3. Vertical section drawing of Cavendish's torsion balance instrument including the building in which it was housed. The large spheres were hung from a frame so they could be rotated into position next to the small spheres by a pulley from outside. This is figure 1 from Cavendish's paper.

the simple expedient of weighing one of the balls to measure the force downward, and applying Newton's law in a situation where G is the only unknown.

In 1797–98, an English aristocrat, Henry Cavendish, carried out a real experiment similar to the one we have mentioned here. It's quite instructive to see how this amateur scientist, who was extremely wealthy, managed to weigh the Earth. His experiment (see fig. 6.3) depended only on the mutual gravitational attraction of four lead spheres: two huge ones and two small ones. He did not need to take account of the gravitational attraction of the Earth. In effect, Cavendish measured the mean density of the Earth, but he is generally credited as the first to measure the gravitational constant.

Cavendish used a torsion balance (which measures a twisting force) in which two small lead spheres attached to a horizontal bar were attracted by gravity to two massive spheres, each weighing 158 kg. The strength of the gravitational attraction twisted the horizontal bar through a small angle, until the twisting force of the thin wire suspending the two small balls balances the force of gravitational attraction. The instrument was protected from wind by enclosing it in a strong box, which was placed in a small external laboratory. Cavendish viewed the torsion wire through a telescope. He found the value of 5.448 times the density of water for the average density of the Earth, giving as we noted a mass of 5.99×10^{24} kg. That is correct to 1 percent and was not improved upon for almost a century.

Now, with both the gravitational constant and the Earth's mass determined as known quantities, we are set up to measure the masses of astronomical bodies. The mathematical details are given in appendix 1, where we show how we could also determine the mass of the Earth by using the period of the Moon's orbit around the Earth, to obtain the same answer as we found above. The Sun's mass (2×10^{30} kg) is then found by using the Earth's orbit around it with the measured Earth-Sun distance and the known period of the Earth's orbit around the Sun, one year. In general, the mass of any large astronomical object (planet, star, galaxy) can be found if there is a smaller object (moon, planet, star) orbiting at a known distance and with a known velocity. If the observational astronomer supplies the velocities and the radii of orbits, then high school algebra allows the mass to be calculated. The method can be applied to planets and their moons; to stars and their planetary systems; to binary stars; to stars orbiting their home galaxy; and so on. Note that these examples are simple systems: one moon orbiting its parent is clearly far easier to analyze than, say, the motions of galaxies in a cluster.

For complex systems, theorists often use the so-called virial theorem, a statistical method of handling the balance between the kinetic energy and the potential energy for an ensemble of orbiting bodies, such as a star cluster or a cluster of galaxies. Although different in detail, the idea behind the method is the same, in principle, as the one that we used for calculating the masses of the familiar local solar system objects—gravity must balance the motions. Later in this chapter we will have more to say about the masses of galaxies, and the means of measuring them.

Before returning to Zwicky and the measurement of galaxy masses in groups and clusters of galaxies, let us turn to what should be the simpler question of how we can we use the method outlined above to measure the mass of an individual galaxy, our Milky Way or one of the neighboring systems.

■ Finding the Mass of the Andromeda Galaxy

Observationally, the quest for galactic masses was undertaken using the Mount Wilson 100-inch telescope and the 36-inch Crossley reflector at the Lick Observatory. We are back in the period when the spectroscopic investigation of galaxies was in its infancy. Nicholas Mayall, a graduate student at Berkeley, California, made an important improvement in instrumentation. While working on his doctoral thesis, he designed for the 36-inch telescope a spectrograph that was optimized for investigating the low surface brightness regions of galaxies. With this new spectrograph, the 36-inch at Lick remained competitive with the 100-inch at Mount Wilson for obtaining the spectra of gas clouds and stars orbiting the centers of nearby galaxies. Everyone realized that the velocities of the orbiting material in the nearby spiral galaxies, as found by spectroscopy, could be used to find the masses of these systems, just as with Newton's laws the velocities of the planets could be used to determine the mass of the Sun. The change in the rotation rate with distance from the center—the "rotation curve"—would tell us how the mass within a sphere of given radius in the galaxy grows as the radius of the sphere becomes larger and larger. It should actually be easier to study a nearby galaxy than to analyze our own Milky Way, since we are embedded in the latter and much of it is obscured from our vision by dust.

Mayall determined masses of the inner parts of nearby galaxies with the 36-inch (the telescope not being big enough to study the faint outer parts); and the results were not surprising. But when in 1937 the young Horace Babcock used the newer Lick spectrograph on the 100-inch telescope to obtain the rotation curve of our neighboring galaxy Andromeda—not just in the central regions but out to more than three degrees (or 125,000 light-years) from its center—the results were surprising.

Babcock had discovered something quite odd. Unlike the situation in our solar system, the velocity of rotation of the galaxy did not fall off with increasing radius. In fact it remained constant or actually seemed to increase in Andromeda's outer, dark parts:

> [T]he obvious interpretation of the nearly constant angular velocity from a radius of 20 minutes of arc outward is that a very great proportion of the mass of the nebula must lie in the outer regions.

Thus, the first serious attempt to determine the distribution of mass within a galaxy already provided a strong hint of the existence of matter that could not be seen, or dark matter. This work of Babcock, along with Zwicky's analysis, constituted the important pre-discoveries of dark matter that were made long before its existence became widely accepted. Both of them were young scientists, working at major observatories. They used the best available observational techniques and applied absolutely standard physical laws, which had been tested from the time of Newton. But somehow, the results were too far from the conventional wisdom to be accepted.

At the time, the outer edge of a galaxy was conventionally defined as the place within which most of the stellar light was emitted, and this gave a mass of perhaps 1.2×10^{11} solar masses for giant spirals in our own nearby cluster, which we call the local group. Since the light within this radius was found to be roughly 2×10^{10} solar luminosities, this gave an overall "mass-to-light ratio" (in solar units) of roughly 6:1. The mass-to-light ratio found in globular clusters or the disc of our galaxy is perhaps 2:1 ranging up to 3:1. Thus there was roughly 2½ times more mass detected for a given amount of light in these spiral galaxies than was found in the nearby and well-measured parts of our own galaxy. The somewhat greater value of the mass-to-light ratio for galaxies as a whole than for the parts within them was certainly considered to be mysterious. But this result was brushed aside as a relatively minor problem, neither serious nor significant. However, the only way that the average value of this ratio could be bigger than the central value was if the value of the ratio in the outer parts was extremely high. These parts must contain a great deal of apparently dark matter but emit almost no light; Babcock's extraordinary discovery of a *very* high mass-to-light ratio in the outer parts of Andromeda was simply ignored.

A personal anecdote may be permitted here. One of us (Ostriker) was asked to give a lecture at the April annual meeting of the U.S. National Academy of Sciences in 1976, summarizing recent discoveries concerning galaxies. Two topics were highlighted—the strange emissions from their central regions (which we now know to be due to embedded giant black holes) and the copious amounts of gravitating mass discovered in the outer parts of these systems. By this year the rapidly accumulating astronomical evidence for dark matter in and around galaxies was in a state that could be summarized for this august audience. The rapt listeners had many questions, but relatively few were from astronomers, who by this point were familiar with the news.

After the talk, in the halls of the imposing National Academy of Sciences building, an elderly gentleman carrying a thick black book approached, asking the speaker if he had time for a few words. It was Horace Babcock, with his doctoral thesis on the Andromeda nebula, written in the year of the speaker's birth, 1937. The speaker had made no reference to Babcock's (forgotten) work, and profuse apologies ensued. In this summary lecture at the NAS, recent radio observations (see fig. 6.4) by Morton Roberts and others had been used to demonstrate how the light of Andromeda was embedded in a much more extensive dark matter halo. Radio and optical work had confirmed each other; an enormous mass of unknown origin surrounded our neighbor, M31.

About three years after Babcock's discovery, Nicholas Mayall, Lawrence Aller, and Arthur Wyse at the Lick Observatory investigated the dynamics of the nearby spiral galaxies M31 and M33. In 1942, Wyse and Mayall noted that the distribution of mass density

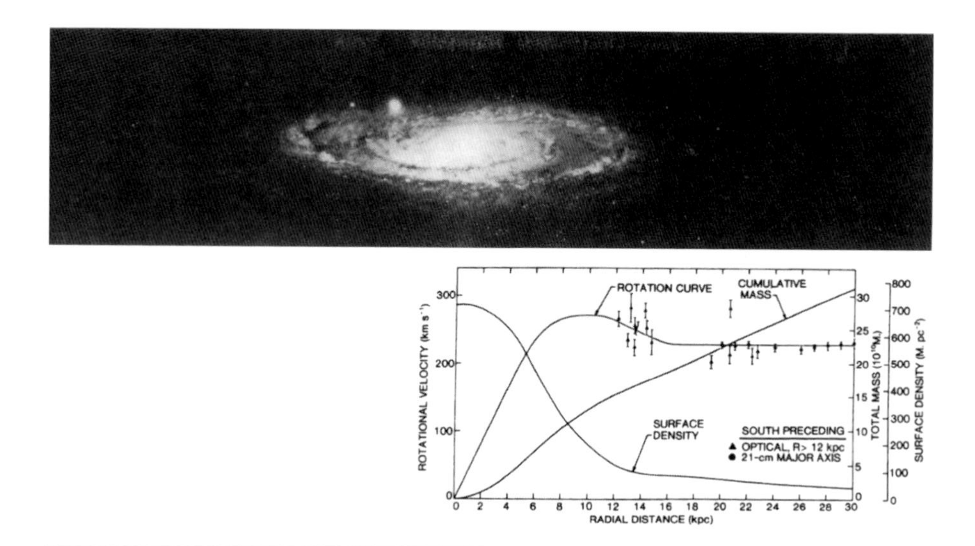

Figure 6.4. Top panel is a picture of our companion galaxy, Andromeda, also known as M31. Below it a diagram showing, on the same scale, the rotation curve and the mass enclosed within a given radial distance. It is clear from this data, obtained from 21-cm radio observations by M. Roberts and colleagues, that most of the gravitationally detected mass is far outside of most of the light, i.e., it is dark. (Ostriker, *Proceedings of the US National Academy of Sciences*, vol. 74, p. 1773, 1977)

in both M31 and M33 implied a large amount of matter in the outer regions of those galaxies, confirming Babcock's result for M31 (the Andromeda galaxy) and extending it to another neighboring galactic system. These results were in complete contrast to the *assumed* density distribution in our own galaxy.

■ Zwicky Finds Dark Matter in Clusters of Galaxies in the 1930s

Now let us return to Fritz Zwicky, whose evidence remained, from 1937 to the present time, as the strongest single argument for the prevalence of dark matter. What did he find, how did he prove it, and why did his work not convince others till decades later? The list of firsts commonly attributed to Zwicky is impressive. He was among the first to realize that (if properly calibrated) supernovae or exploding stars could be used as standard candles for cosmological work. He hypothesized that immensely dense neutron stars would be formed in these giant explosions. And he was the first to realize that normal galaxies could act (following Einstein's earlier work) as gravitational lenses producing multiple images of background objects.

Each one of these predictions was confirmed years or decades later. But, despite these demonstrations of brilliant prescience, Zwicky never cast off the reputation of being a crank. Zwicky had first written on how to use virial methods (which statistically find the gravitating mass needed to contain the observed motions) from classical mechanics for determining masses of astronomical bodies in 1933. Then, in 1937, his paper *On the Masses of Nebulae and Clusters of Nebulae* used the virial theorem for the first time to measure galactic masses. He showed that the typical mass is roughly one hundred times higher than one would have estimated from the light emitted by these galaxies. Finally, he speculated on the nature of the extra matter and proposed definitive tests for his bold speculation.

Zwicky had started in a conventional astronomical fashion, as we noted in chapter 5, working on sky surveys, cataloging the objects that he saw in the sky, and had found in the northern skies what seemed to be clusters of galaxies. The Coma cluster, whose central parts are shown in figure 6.1, is the paradigmatic example. These were more or less spherical assemblages of normal stellar systems, but they were immense, each one extending over nearly three million light-years and containing roughly one thousand individual galaxies. Many such clusters were found and they had structures similar to clusters of stars, but of course on a much grander scale. In the case of star clusters we have multiple lines of evidence that the stars therein are all kin; that is, they have a common origin. They have very similar chemical compositions, nearly the same ages, and they all orbit around the centers of the systems in complete accordance with the expectations from Newton's laws. The simplest application of Newton's law of gravity would be to look at a star in a circular orbit in the outer parts of the cluster, measure its velocity and the diameter of the orbit, and then calculate the mass of the cluster in exactly the manner described in appendix 2.

It is clear from simple physical principles that if, instead of a single star in the outer parts of the cluster, we used the *average* of the (squared) velocity of all the cluster stars, there should be a similar equation. For a system in equilibrium (neither expanding nor contracting significantly) one can show that the gravitational binding energy must be exactly twice the kinetic energy of the motions of the stars: this is the virial theorem and it will always hold true for a system in equilibrium that is neither expanding nor collapsing. When this method is applied to star clusters, the derived total mass is in good agreement with the mass estimated by simply inventorying the stars and adding up the mass from those in the various categories. The virial method gives the expected answer.

When Zwicky turned his attention to the dynamics of the nearby and very luminous Coma cluster of galaxies, he used the same method and found that the large spread of internal velocities (~1000 km per second) implied a gigantic mass (roughly 10^{15} solar masses), far greater than one would have estimated from the number of galaxies and the light emitted by them. He quoted a mass-to-light ratio of 500:1, which is much higher than the ratio astronomers were obtaining for individual galaxies. If we put modern refinements into the calculation, we still obtain ~300:1 for this ratio. This is more than one hundred times the value for the same ratio in star clusters and fifty times the value estimated for the visible parts of spiral galaxies.

Even allowing for the fact that galaxies in clusters are primarily elliptical galaxies, which do not produce as much light per unit mass as the spiral galaxies, the discrepancy remained very large, so large, indeed, that Zwicky was led to propose other methods for weighing the galaxies in clusters and to consider the possibility that the mass in stars *between* the galaxies might be throwing off the results.

Clearly he was somewhat unnerved by his own discovery. He presented these results cautiously; and they were not given much credence by other astronomers. The curious citation history of his 1937 paper shows this neglect. In the first forty years after its publication, only ten papers referred to this work, but there were twenty-three citations in 2009 and a still higher forty in 2010. The steady climb in citations of the work by other scientists began after it could be placed in the context of the growth of structure in a universe dominated by dark matter. Only then were cosmologists able to grapple with the revolutionary nature of the evidence from the Coma cluster and other similar groupings; we now know that rich clusters of galaxies are the most massive self-gravitating objects in the universe. In 1937, Zwicky named the strange stuff that he had found *dunkle Materie*: dark matter. The expressive moniker was all but forgotten until its revival some sixty years later.

Research on and debate about the masses of galaxies lay fallow for decades. There were many papers on the rotation curves of normal galaxies, including a series by Margaret and Geoffrey Burbidge and their colleagues, and these gave essentially the conventional answers, but they tended to focus on the inner, optically bright parts of the observed systems (see fig. 6.6 *top*), where it was not easy to distinguish between galaxies with or without massive halos. However, there were several other clues accumulating which indicated that outside of the bright parts of galaxies there was a great deal more matter—enough to be consistent with the strange observations of Zwicky and Babcock decades earlier.

■ The Rediscovery of Dark Matter in the 1970s

In 1974 one of us (Ostriker), together with Jim Peebles and Amos Yahil, wrote a paper putting these clues together, fulsomely entitled *The Size and Mass of Galaxies, and the Mass of the Universe.* That paper begins:

> There are reasons, increasing in number and quality, to believe that the masses of ordinary galaxies have been underestimated by a factor of 10 or more. Since the mean density of

> the universe is computed by multiplying the observed number density of galaxies by the typical mass per galaxy, the mean mass density of the Universe would have been underestimated by the same factor.

The excitement of the moment is palpable. The paper presents no original calculations nor, in fact, any new observations. What it does is summarize the strong evidence, as it stood at that time, for the concept that there was a great deal of gravitating matter far from the centers of normal galaxies. This matter is located in galactic halos, the outermost regions that emitted relatively little starlight.

Ostriker and Peebles had written a paper the previous year, 1973, entitled *A Numerical Study of the Stability of Flattened Galaxies: or, Can Cold Galaxies Survive?* which did contain new calculations. That paper had shown that our own galaxy and others like it, which seem to consist mainly of a flat disk of stars, are actually embedded in a more or less spherical component of similar or greater mass. They showed that if this were *not* the case, the galaxy would have become wildly unstable, ultimately forming into the shape of a giant bar, which has obviously not happened, since our galaxy contains at most only a modest central bar. This instability could not have been present, and so the mass of our galaxy must be dominated by a fairly spherical component. But this is not seen in the distribution of the visible stars. The observed stars in our galaxy and in equivalent ones show only a modest central bulge, and this quasi-spherical component contains far less than one-tenth of the total starlight in the overall system. This argues for the probable existence of an additional massive and invisible dark spherical system *within* the galaxy:

> If our Galaxy does not have a substantial unobserved mass in a hot disk component, then apparently the halo (spherical) mass interior to the disk must be comparable to the disk mass. Thus normalized, the halo masses of our Galaxy and of other spiral galaxies exterior to the observed disks may be extremely large.

Here the authors were arguing that, if the dark matter inside of a galaxy (that is, within the region emitting abundant starlight) was substantial, then it seemed likely that outside of this region the ratio of dark to light emitting matter would be still larger. This indirect line of reasoning was based on a theoretical argument of what must exist in order to prevent an unobserved instability and, by itself, would not have been very persuasive to the "show me the evidence" mind-set of the sensible astronomical establishment. But the second paper (1974) did have the evidence (see fig. 6.5). Here is what it showed.

Moving from the inside outward, the paper showed first the conventional result that, within the disk of our galaxy, the rotation curve was approximately flat: the rotational velocity did not decrease as larger and larger orbits were examined. This implied that the mass within those orbits must increase in proportion to the distance from the center of the galaxy. The light within our galaxy and others is concentrated toward the center, so it was expected that the mass would be concentrated there as well. In the solar system, the orbital velocities decrease with increasing distance from the Sun as described in Kepler's laws. The velocities within our galaxy did not show that pattern. When this unexpected local flatness of the galactic rotation curve became a known fact, it was thought to be slightly strange but not impossible, because the enclosed mass does increase outward due to the numerous disk stars and spiral arms of our Milky Way.

Farther out, on the periphery of the galaxy, there are small galaxies and globular clusters that are companions to the Milky Way. These have sharp edges to their light distributions, which were interpreted as tidal cutoffs, the boundaries where the pull from our giant galaxy would overcome the self-gravity of these small systems and tear off the outer stars. The position of the tidal cutoff could thus be used to measure the mass of the galaxy (interior to the radius) at which the small system was orbiting. And a rather large mass was also found by this method.

Finally Ostriker, Peebles, and Yahil looked out still farther, reviewing a simple argument that had been given by Franz Kahn and

Ludwig Woltjer in 1959. Those authors noted the basic fact that our companion galaxy, the Andromeda galaxy, was coming toward us at the quite substantial velocity of 300 kilometers per second. It is one of the few galaxies that is approaching us rather than receding from us. How could this happen in a universe where almost all galaxies are flying apart from one another? The culprit must be gravity.

By way of explanation, they suggested that our galaxy and Andromeda galaxy started out with velocities that satisfied the Hubble flow (moving away from each other), but the gravitational attraction between them was so great as to reverse the motion and send them hurtling back toward one another. Following this suggestion, Ostriker, Peebles, and Yahil computed that the mass must have been 1,000 billion (10^{12}) solar masses, far more than could easily be accounted for by the observable stars! Others (including Jim Gunn and Jan Oort) had checked the measurement and calculations and arrived at essentially the same result. In fact, many authors had looked at other nearby binary galaxies and small groups and found the same disturbing result: masses of 10^{12} solar masses per galaxy were typically required to account for the observed motions, far greater than could easily be accounted for by the visible stars.

Then, when all of the measurements of the mass of our Milky Way to greater and greater distances from the center were put onto a single plot, they found that the mass enclosed within a sphere seemed to follow in an uncanny fashion a simple rule: it was proportional to the radius of the sphere out to radii of about 1 million light-years, where it reached the value 10^{12} solar masses.

This result was considered extremely strange, because photographs show normal galaxies (and our own galaxy as well) to be rather concentrated with regard to light (see fig. 6.4). The central parts are bright and then the brightness falls off rapidly with distance. Most of the light is concentrated in the inner 10 kpc. But here was evidence, and it was the best that was obtainable, that the mass distribution of the galaxy was completely different from that expected from the observed distribution of the starlight. Most of the mass was more than 100 kpc from the center, where, for all

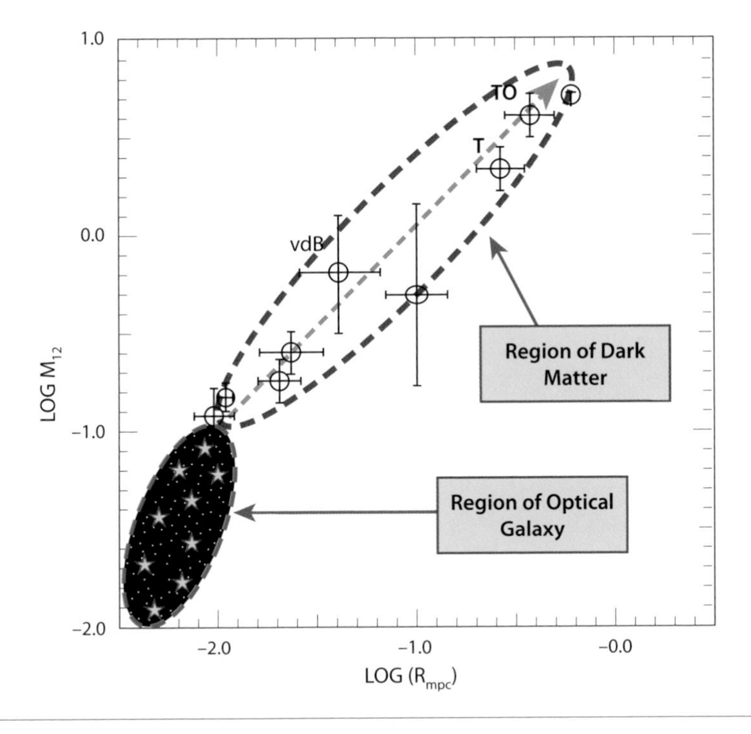

Figure 6.5. The measured mass, *M*, of our galaxy within a distance, *R*, of its center. A dramatic increase of the measured, enclosed mass far outside of the observed optical galaxy was found—giving strong evidence that our galaxy was embedded in a dark matter halo. (Adapted from J. P. Ostriker, P.J.E. Peebles, and A. Yahil, *Astrophysical Journal*, 193 L1 (1974)

intents and purposes, galaxies (including our own) were dark. So the light seemed to be concentrated in an inner region surrounded by a massive halo containing ten times more mass than was associated with the stars in this inner region. Most peculiar.

Since these results were often obtained using the dynamical law called the virial theorem, many authors referred to the unwelcome result as the *virial discrepancy*, as if the problem was with the application of Newton's laws! In 1974, Ostriker, Peebles, and Yahil reviewed the literature on rotation curves of other galaxies and found that many astronomers had found flat rotation curves out to large distances from the centers of the studied galaxies, all indicating that the mass in every one of these systems seemed to increase in the outer parts at a rate proportional to the size of the region studied.

Another way of expressing these peculiar results was to look at the ratio of light output of a galaxy divided by its mass, the "mass-to-light ratio." Defining units so that this ratio is 1:1 for the Sun, we noted earlier that astronomers had found that the subsystems within our galaxy had mass-to-light ratios like that of the Sun—approximating 2:1. But it seemed that when they measured the total light and the total mass out to a large radius, the measured mass-to-light ratios were greater than 100:1. What was the nature of the extra mass in the outer parts of galaxies? People, especially popular science writers, called it the "missing matter," but the problem was not that matter was missing, rather, that it was there in abundance! It was the light that was missing in the outer parts of galaxies.

All of this work started with and was based on the familiar spiral galaxies. However, they are in a minority with regard to mass: what about the numerous and typically more massive elliptical galaxies, which display many fewer features than spirals? Herbert Rood, writing with several others in 1974, reviewed the dynamical evidence for these systems. They found even bigger mass-to-light ratios, up to about 200:1 in solar units. Thus, the accumulating modern evidence was really quite consistent with the results found by Zwicky and Babcock in the 1930s, but the picture was now clearer. Each normal galaxy has within its optical image both ordinary matter and dark matter in roughly comparable amounts. This explains the factor of two discrepancy found earlier between the expected and observed mass-to-light ratios within galaxies, and it is consistent with the dynamical argument in the first paper by Ostriker and Peebles. But then, surrounding each galaxy, there is a halo of predominantly dark matter containing roughly ten times as much mass as the combined stellar and dark matter mass interior to the visible system.

Given that most of the light in the universe came from either elliptical or spiral galaxies, Ostriker, Peebles, and Yahil now felt that they could weigh the cosmos by the simple method of multiplying the light in the typical volume of space (big enough to contain both galaxies and voids) by the total mass-to-light ratio of the galaxies in that volume, allowing for both the inner and outer components of each galaxy. The mean density of light in the universe (the amount of light emitted per unit volume on average) was fairly well known simply from the surveys of light-emitting galaxies. They obtained a number by this method that was roughly ten times higher than the number generally accepted at the time. In the abstract of the paper they boldly asserted:

> [W]e determine the local mean cosmological mass density $\approx 2 \times 10^{-30}$ g cm^{-3} corresponding to $\Omega_{matter} = \rho/\rho_{crit} = 0.2$. The uncertainty in this result is not less than a factor of 3.

Recall that in chapter 3 and in appendix 1, we introduced the Greek letter, capital omega, Ω_{matter} that cosmologists use to designate the total density of matter (in all forms added together). If it has the value of unity or less, then gravity is not strong enough to prevent the universe from expanding forever. The result that Ostriker and his colleagues determined by using this elementary method was $\Omega_{matter} = 0.2$, which was less than the magic value of unity, but it was roughly ten times greater than expectations that ignored the dark matter. Also, their estimate turned out to be amazingly close to the best current estimate, 0.27; this was good fortune. And it was rather exciting. All of the evidence for huge amounts of dark matter did hang together, but it was a great deal for the astronomical world to swallow. It took time and yet more evidence to convince most of the astronomers.

■ Rotation Curves Confirm the Case for Dark Matter

A little later, in the late 1970s and the early 1980s, the most persuasive additional evidence came forward from the work of Vera Rubin, Kent Ford of the Carnegie Institution in Washington, and their collaborators. Working with the recently invented image tube spectrograph they provided the observations that were taken as definitive. After their work was published, most of the remaining skepticism concerning dark matter evaporated. This group used new electronic technology, which enabled them to measure velocities farther from the galactic centers than others had achieved, showed case after case of giant spiral galaxies having flat rotation curves to quite large distances. Figure 6.6 *bottom* shows data from their paper of 1980. This was the norm; the case for extended mass distributions in normal spirals was established in a definitive fashion by the late 1980s by Rubin and other groups, so the existence of dark matter in the outskirts of spiral galaxies became a proven fact. The Rubin-Ford results were not different from what Babcock had found to be true for our neighbor Andromeda. And the evidence was in fact consistent with the less observationally persuasive results of other radio and optical observers that we have referred to. But it was the accumulation of evidence, the fact that flat rotation curves were the norm, and were clearly quite extended to far beyond the optical images of the galaxies, which proved persuasive to all.

There is a further curious twist to this history that we noted previously. Earlier work in the late 1960s by Margaret and Geoffrey Burbidge had already shown rather flat rotation curves for many nearby galaxies. Their data were not as good as that later obtained by Rubin and her co-workers, and fitted to reasonable accuracy quite a range of different rotation curves including (though the fit was poor) curves that had the decline (expected at that time) in the outer parts of galaxies (see fig. 6.6 *top*). This allowed the Burbidges to draw the conclusions expected at that time, of conventional mass estimates, even though, to the modern eye, their data clearly show the flat rotation curves that we have come to know and love. But, since the instruments available when this work was done were not sensitive enough to measure far out from the centers of the galaxies, they did not probe the domain where the dark matter dominated and so, even if they found flat rotation curves, they did not reach out to measure very large total masses. This is a reminder that occasionally observers can be unconsciously biased by what they expect to find.

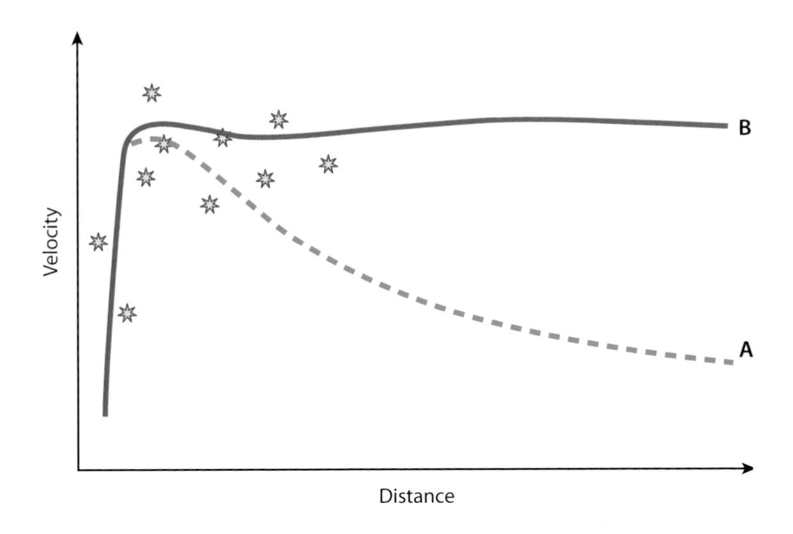

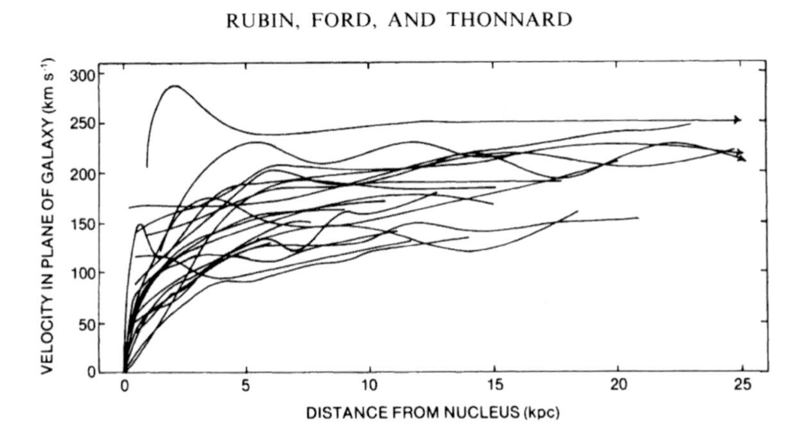

Figure 6.6. The top panel diagram shows the rotation curve of a typical spiral galaxy: (A) expected from the distribution of the light and (B) from modern observations. Spectroscopic data obtained by Margaret and Geoffrey Burbidge in the 1960s, indicated schematically by the blue stars, could not discriminate between the two types of curve. The difference between the curves is attributed to dark matter. B is described as a "flat rotation curve." The bottom panel shows a summary of data obtained by V. Rubin, K. Ford, and N. Thonnard for late type spirals. (*Astrophysical Journal, 238: 471, 1980*)

As late as 1961, a summary by three leading astronomers (J. Neyman, T. Page, and E. Scott) at a large "Galaxy Symposium" conference was at a loss to explain the excessive masses indicated for galaxies in clusters (by Zwicky), taking seriously the possibility that they were exploding or were merely chance accumulations of matter. The simple explanation that astronomers had greatly underestimated the masses of galaxies was unthinkable at that time. But, gradually, by roughly 1980 the consensus that galaxies contained unseen matter became broadly accepted. All the data from the motions of stars in galaxies or from galaxies in groups and clusters were realized to be consistent with having some dark component of the universe with ten times more mass in it than could be accounted for by the visible stars and five times more mass than in all of the ordinary matter.

This revolution in thinking about the contents of the universe had taken half a century to be widely accepted since the first neglected papers by Zwicky and Babcock, but only a decade since the problem began to be studied in earnest in the early to mid-1970s. Now, in order to be taken seriously, all cosmological models had to contain, as the majority element, dark matter.

The dark matter was not distributed smoothly in the universe. It appeared to exist in the general vicinity of the stellar concentrations, but it was less concentrated than the stars. It preferentially populated the outer parts of galaxies. Admittedly, it was strange stuff, but it is important to note that the *kind* of measurements that were used to detect dark matter and to calibrate the amount of dark matter were not types of measurement that were in any way special. They used the same methods and the same laws of Newton that we adopt to find the masses of the planets and our own Sun.

Skeptics, independent thinkers, and theorists with a taste for risky analysis have said, in their different ways, that perhaps gravity does not follow exactly the inverse square law, and instead becomes stronger at large distances. Or, maybe gravity has an explicit time dependence and was different at earlier times. Or, is there some strange extra interaction that affects gravitating masses? If the evidence for dark matter were only what has been presented so far, then human ingenuity and the desire for fame would surely have led individual scientists to invent some other way to explain the evidence that did not invoke dark matter. In fact, there have been many attempts to do just this. However, none of them is able (in the view of the authors of this book) to explain all of the evidence that supports the hypothesis of dark matter. These theories may well pass the tests that they were designed to pass, but they fail some of the other tests. We are reminded of the pre-Copernican philosophers and their attempts "to save the appearances."

■ More Recent Multiple Lines of Evidence for Dark Matter

Let's now look at some of the other lines of evidence that point to the existence of dark matter. Can we see dark matter in action? Yes we can. There is a rare and wonderful example of a system in which the ordinary matter and the dark matter have been separated from one another. In the direction of the constellation Carina, a distant pair of clusters of galaxies seem to be colliding in a high-velocity impact.

It is called, for obvious reasons, the bullet cluster. As in normal clusters, most of the ordinary (or baryonic, atomic) matter is primarily in the form of gas seen between the stars. The simple explanation for this is that the process of condensing gas into stars seems to be inefficient, with only perhaps one-tenth of the gas being so converted. In a collision between two giant systems, the stars would almost always pass by one another without individual stellar collisions and would emerge as separate systems (see the blue patches) after one cluster had passed though the other one. That is exactly what is seen. However, the more abundant gas atoms in each system cannot help colliding with their counterparts in the other system at the moment of encounter, and they will thus be left halfway in between the two emerging stellar systems—and that is also exactly what is seen (see the red patches).

Figure 6.7. The Bullet cluster in Carina consists of two colliding clusters of galaxies. The discovery of this cluster in 2004 was equivalent to the discovery of the Rosetta Stone, in that the motions of the galaxies could not be understood in terms of the normal matter seen by X-ray telescopes but easily understood in terms of non-interacting dark matter. Red shows the gas and blue shows the matter distribution as determined by gravitational lensing. (Image Credit: X-ray: NASA/CXC/M. Markevitch et al. Optical: NASA/STScI; Magellan/U.Arizona/D. Clowe et al.)

Now, let us ask about gravity. First let us consider the possibility that there is no dark matter. If all the gravity is due to the ordinary matter, then the source of gravity must be gas in between the two stellar lumps (since in terms of mass, the gas dominates over the stars by a factor of ten), and the direction of gravitational force would point to the spot halfway between the two stellar systems where the gas is concentrated. But if dark matter (made of particles that rarely collide with one another) predominates, then it acts like the stellar components and will emerge along with the stars in two separate lumps, and the gravitational force would point to the centers of the two stellar systems. By looking at the orbital velocities of the stars, one can easily determine the direction of gravity, and observers found that it is clearly pointing toward the centers of the blue patches, the centers of the two stellar systems, and not toward the gas in between them. This is a clean case: a Rosetta Stone for dark matter! As far as we know, none of the rival theories can explain the bullet cluster.

Still another test for the reality of dark matter is through gravitational lensing that we introduced in chapter 1. Zwicky pointed out in 1937 that galaxies and clusters of galaxies would be much more efficient gravitational telescopes than are stars. The separation of the images produced by these massive systems would be so wide that they should be clearly detectable and not be obscured by the intervening galaxy or cluster that acts as the lens. And in fact, many years later it has been amply observed in hundreds of cases. Clusters of galaxies do produce multiple images of background objects in just the predicted manner and with just the mass expected from the sum of the normal matter and the accompanying dark matter.

The distribution of these images (with a good example, cluster Abell 2218, shown in figure 1.2) has been studied in careful detail, and it agrees quite nicely with what would be expected from the gravity exerted in a system dominated by dark matter. The dark matter here acts exactly as it should, causing light to bend according to Einstein's laws as well as inducing the motions of the galaxies following Newton's laws. And the amount of dark matter found in clusters using the gravitational lensing method is exactly the same as the amount found by Zwicky using the motions of the galaxies in clusters and the virial theorem. In our illustration of a cluster of galaxies earlier in this chapter, the obvious arcs are in fact distorted images of background galaxies and they come in pairs seen on opposite sides of the intervening cluster.

One final example of how the expectations arising from the existence of dark matter were verified involves the growth of structure in the universe. We introduced this subject in chapter 5. The universe that we see around us at the current epoch is, as we have noted, lumpy in the extreme and made up of stars, gas clouds, galaxies, clusters of galaxies, and vast empty voids. But we think that the Big Bang ended with a fairly uniform distribution of matter. This is a supposition, but it was confirmed by the early measurements of the cosmic background radiation, which, during the decades of study of the cosmic background radiation from the ground, seemed to show the sky to be quite uniform. How could this be? How could structure grow from uniformity? Impossible. There must have been some "seeds" of the later structure in the early, apparently featureless, background.

We noted in the last chapter that theorists such as Peebles in the United States and Zeldovich in Russia insisted that these seeds *must* be there in the cosmic background radiation and that the observers should design more sensitive experiments to find them. As we have seen, in 1991 the COsmic Background Explorer satellite was launched with, at last, instruments sufficiently sensitive to detect the low-level, early fluctuations and the overall radiation spectrum. As mentioned in chapter 5, the Princeton colloquium announcing the first results included a flash preview of the mapping experiment, the detailed results of which were completed only a year later (see fig. 5.5 *top*). The Princeton audience immediately grasped the potential of the preliminary map: *the predicted seed fluctuations in the cosmic background radiation had been detected!*

The fluctuations seen were very low in amplitude (about one part in 100,000 around the mean). Could these have grown in the available time to what we see today if what we see is all there is? Alas, the answer was no. The visible matter in galaxies fails by a substantial amount to provide enough gravitational perturbations to grow to what we see around us. They fail by much more than the likely errors in the observations or in the calculations. Could the extra gas between the galaxies add enough mass to do the job? No. We knew from the light element nucleosynthesis arguments (chapter 4) and from direct observation that there was not enough in either gas or stars. What could explain it?

Then, in the room where the pictures were first shown, someone asked, "But what if there is also dark matter? And the dark matter fluctuations, even though invisible, would have exerted enough extra gravitational force to help the perturbations to grow;

would that solve the problem?" The answer was a resounding "yes!" Calculations by Peebles and others showed just how this could happen. And what amount of dark matter was needed? Just the same ratio of dark matter to ordinary matter (five to ten times more dark matter than visible matter), as was seen nearby in galaxies and clusters. Since that early time, the observations of the background radiation have become much more accurate, and the precision of the calculations has vastly improved. At present the best estimate for the mass density in dark matter now comes from this source. The seven-year data from the WMAP satellite launched in June 2001 gives $\Omega_{\text{dark}} = 0.23 \pm 0.01$ and $\Omega_{\text{matter}} = 0.27 \pm 0.01$, quite close, as noted earlier, to the first estimate from Ostriker, Peebles, and Yahil in 1974 of $\Omega_m \sim 0.2$. The amount of matter in the universe, both dark matter and ordinary matter, is about one-quarter of the critical value, the value at which gravity and expansion just balance.

There are still other independent lines of evidence. But by now the reader has the flavor of the historical sequence. What started out three-quarters of a century ago as an apparently mad conjecture by an eccentric astronomer (and was ignored for almost four decades) has now been tested and confirmed by numerous different methods and innumerable observational tests; and they all conspire to give essentially the same answer. There really is some strange substance, which we call dark matter, which acts gravitationally just like ordinary matter, but does not seem to interact either with light, or with itself, or with normal matter in any other way except through gravity. It has been present since (at least) the cosmic background radiation was formed, and it is roughly six times more abundant than ordinary matter. But what is it *really*? The answer to that important question is easy! We do not know. Later we will discuss current experiments and conjectures on this central issue, but, at the end of the (current) day, the answer remains: the nature of dark matter is a mystery.

Chapter Seven

Dark Energy—or Einstein's Greatest Blunder

■ A Curious Situation

"Curiouser and curiouser!" Cried Alice (she was so much surprised, that for the moment she quite forgot how to speak good English.)

We start this chapter with a brief quotation from Lewis Carroll's *Alice's Adventures in Wonderland*. This famous book is not just a novel for young children. It is a masterpiece of imaginative literature, with rich symbolism, coded language, nonsensical conversations, and poetry. The Reverend Charles Lutwidge Dodson was a mathematician and lecturer at Christ Church, Oxford, and, while he and Alice were not talking of cosmology, they might have been. We hope that you are becoming comfortable with the curious dark matter, but our tale is about to become still stranger, as the last vital ingredient of the universe is introduced, and this additional component has no counterpart on Earth. It is a very weak force that, unlike any other, becomes stronger with increasing distance.

In the previous two chapters we have explored at some length the interaction between ordinary matter, dark matter, and the force of gravity. We have seen that dark matter is more abundant than normal matter, and we have developed the concept that any attempts to understand how structure arose in the universe depend on the dominance of dark matter. We also tried to make clear that the introduction of dark matter was not just adding a patch to a tattered theory. On the contrary, there were signs pointing to it from several independent directions, and often these signs, bright and clear as they were, had been overlooked until, slowly, during the 1970s and '80s, the case became so overwhelming that the concept was universally accepted.

But we have had little to say about energy, apart from the energy of the microwave background radiation. We have established that the evolution of the expanding universe, as well as the growth of structure within it, are related to the strength of gravitational fields, and the potential energy that they contain. In this chapter we will give deeper consideration to the role of energy in the expanding universe. We will indeed see that the more we explore the universe (Wonderland), the more we seem to uncover a sequence of intellectual puzzles that sometimes seem like nonsense when we first encounter them. Let us begin with a fundamental question provoked by Newton's physics and left unsolved for centuries: with gravity the strongest force on cosmic scales, what restrains the universe from collapse?

■ Will Gravity Lead to a Collapse of the Solar System?

Newton's universal theory of gravitation had incorporated an obvious potential flaw: since gravity is always attractive, pulling things together, the solar system (which *was* the universe so far as Newton was concerned) might, in the long term, collapse under its own weight so to speak. As we have seen, Newton (a devout Christian) assumed a "hand of God" approach as a fix for the requirement of stability. But other scientists' fears that there might be orbital instabilities lurking in Newton's clockwork universe led to more and more refined dynamical investigations.

For instance, Pierre-Simon Laplace spent many years, from about 1773, studying the stability of the solar system through careful analysis. Since his achievement is highly technical, we will give just one example of his analysis of the worrisome potential flaw in Newton's theory. Newton assumed that the speed at which gravitational forces propagate through space was infinite. In Newtonian physics, if one of Jupiter's moons crosses from one side of Jupiter to another, the rest of the solar system knows about this tiny change of gravitational forces instantly. That is of course impossible, we now know, since nothing can propagate at infinite velocity; the upper limit at which any signal can move is the speed of light.

Laplace investigated the effect of gravity propagating at a finite (though still very high) velocity. He found that at large distances a finite speed for gravity would be akin to the actual force of gravity being a little weaker than in Newtonian theory, and allowing for a finite propagation velocity did not cause the theory to collapse. This work is flawed, but at least Laplace made the problem rational by expelling Newton's God of the gaps (in knowledge) from the solar system. The stability of the solar system was established, although Laplace did find a slight problem with Mercury's orbit that was left for later study.

There the matter might have rested for centuries, but for Urbain Le Verrier, who followed his famous triumph of 1846, the prediction of the position of Neptune, with the unwelcome announcement in 1859 that, with the good data that he had, the planet Mercury seemed, in fact, to be seriously misbehaving. Le Verrier had analyzed observations from 1697 to 1848, and discovered that the rate of precession (the rate at which the whole elliptical orbit rotates) was twice that predicted by Newtonian gravitation. Another way of visualizing precession is that each time the planet loops the loop, so to speak, it overshoots and misses a perfect ellipse by a tiny amount. Newton's theory could handle this errant orbit, but not quite, failing by a factor of two in predicting the rate of precession. To save the appearances (as the ancients would have thought of it), mathematicians tinkered with celestial mechanics in an ad hoc fashion, without achieving anything useful or resolving the puzzle. Apparently there really was an error somewhere. Perhaps Newton was somehow wrong, but it was at least reassuring that Le

Verrier still found no catastrophic instabilities that would cause a future collapse of the solar system.

There was no instability, but it really became clear that Laplace and Le Verrier had found a problem. Newton's laws were inaccurate in predicting the orbit of Mercury. It was only in the mid-1960s that radar ranging of Mercury was sufficiently precise to prove, beyond all doubt, that Einstein's theory addressed the error and computed this planet's orbit correctly. General relativity theory passed this very strong test with flying colors.

However, the overall perplexing issue of how we can have gravity—the single dominant force on large scales—always pulling things together, but nevertheless we still see an apparently stable universe of stars, would not go away. The twentieth-century work by Harlow Shapley on the Milky Way, described in chapter 2, showed that we lived in a rotating galaxy prevented from collapsing inward by centrifugal forces, in the same fashion that the solar system staves off collapse via rotation. Thus, as our concept of the size of the universe expanded, the problem of apparent stability in the face of gravity remained unsolved on the largest scales; it was pushed out further still to the realm of the galaxies and was left for future work. The good behavior of the universe in refraining from gravitational collapse remained as a fundamental issue to be faced somehow by either philosophy or science. We will first review what was expected; an expanding universe is either moving fast enough to overcome its own self-gravity, or it must re-collapse.

■ Expected and Unexpected Motions of Thrown Stones and Hubble's Universe

Let us start with one of Einstein's tools: thought experiments. Elementary physics can tell us what to expect in an expanding universe. In a simple exercise in chapter 3, with the details spelled out in appendix 1, we followed the evolution of a tiny spherical piece of the universe. You may remember that we found that there are only two possibilities in the standard model of an expanding universe. Either gravity will win and the little sphere will reach a maximum size and then re-collapse, or gravity will at first slow its expansion, but ultimately lose the battle and ultimately be so weak that after a long enough time the little sphere will continue to expand at a constant velocity. These are the two natural possibilities for our toy model. However, reality has confounded our expectations. Neither possibility is now believed to be correct in the universe around us. That is the puzzle we explore in this chapter.

Here's yet another thought experiment: we ask what happens when you throw a stone straight up into the air. At first the stone is moving rapidly (the velocity of your hand when you let it go), and then it slows and slows, until its motion stops, and then it falls back to Earth faster and faster until it hits the ground near you. With what velocity does it hit? The same velocity it had when you let it go, but in the opposite direction. And suppose you threw the stone harder and faster, what would happen? We all know the answer. It would simply go higher and on return hit the Earth going faster—hitting with a bigger thud. If you threw it fast enough, could you get the stone to fly off the Earth? The ancients would have thought that this was a strange but not a crazy question. No one is strong enough to perform this experiment, but we do know that we can attach a rocket to the stone and give it a large enough velocity to leave the Earth, as this is how we have sent probes into space that leave the Earth and travel to distant planets.

About how fast would we have to throw the stone to make it leave permanently? That velocity is called (naturally enough) the escape velocity and is the same anywhere on the planet. Let us first try for a rough estimate, of the kind that the ancient Greek philosophers or experimentalists might have made, and certainly Galileo could have made. When a stone is dropped from a height, it falls faster and faster with increasing time, reaching higher and higher speeds, if dropped from greater and greater heights. The velocity on hitting the ground increases with the height from which the stone is released. So far everything is obvious.

Can we be more precise? Our falling stone is being accelerated by the gravitational attraction of the Earth. The entire planet is pulling on the stone. The acceleration due to the gravity of the Earth is 9.81 meters per second per second. If a massive stone that falls from the top of a tall building takes 5 seconds to hit the ground, it does so at a final speed on impact of 50 meters per second. How high is the building? Well, since the average velocity is 25 meters per second and the fall took 5 seconds, the building must be 125 meters in height.

Let us now try to work out the details. The final velocity, we'll call it V, is given by the acceleration g multiplied by time to fall, t: $V = gt$. The distance traveled, d, is the average velocity, $\frac{1}{2}V$, multiplied by the time: $d = \frac{1}{2}Vt = \frac{1}{2}gt^2$. In this last equation we put $t = V/g$ from the prior equation, rearrange it, and arrive at $V^2 = 2gd$. The final impact velocity is proportional to the square root of the height of the building.

Our next step is to time reverse all of this and imagine the stone thrown upward. How fast would we have to throw the stone to get it to leave Earth for good? Just suppose d, the distance thrown, were as large as the radius of the Earth—we threw it up one earth radius high or higher. Then, by Newton's law the force pulling it down would be so much weaker that it might escape altogether. The force acting, per unit mass, is $g = \mathrm{G}\, M_e/R_e^2$. Substituting this result for g in the previous equation we now have $V^2 = 2\, \mathrm{G}\, M_e/R_e$ for the required velocity.Putting in the actual values of the constant G, and the mass and radius of the Earth ($M_{e,}$ R_e), we can show by ordinary arithmetic that V has the value 11.2 kilometers per second (about 25,000 miles per hour). This is in fact the escape velocity from Earth, the velocity at which a stone would have to be thrown straight up, to escape the Earth's gravity. Now suppose that the rocket launched the stone at 30,000 miles per hour; then the stone would slow down by 25,000 miles per hour and finally would leave our planet still traveling at a speed of 5,000 miles per hour, sailing out into the solar system at that speed. Once out in the solar system it would face the gravitational forces of the Sun and the other planets. This is what we expect. It is what happens when we launch rockets.

Let us continue our thought experiments by considering the fate of *Voyager 1*, a NASA spacecraft launched on September 5, 1977. It has left our solar system, and has traveled farther than any other object launched from Earth. Its velocity is about 38,000 miles per hour, and it is on an extended mission that should last until at least 2020 because it is still in radio contact.

The spacecraft is racing away at nearly a million miles a day. But now suppose that, after breezing along at this rate for a few years, the mission controllers noticed that, little by little, the velocity (with respect to the solar system) was *increasing*, despite the fact that the spacecraft is light-years away from stars other than the Sun. The farther away the spacecraft was, the faster it went, receding from us. That behavior would be strange indeed! Curiouser and curiouser, as Alice puts it. Such a discovery would be perplexing, worse than dark matter, much worse. Dark matter was discovered through the now familiar gravitational force that Newton introduced, but if the observation we just described were done, it would be due to some exceedingly peculiar new force, for which we have no examples on Earth.

The astonishing fact is that the behavior we have just described as a thought experiment is actually occurring on a gigantic cosmological scale. There seems to be a force, which we do not understand at all, that pushes galaxies and clusters of galaxies apart from one another. It opposes gravity, which, like any reasonable force, weakens with increasing distance, and instead it mysteriously *increases* in strength proportional to the distance between the galaxies. It is acting now in the universe, at our own epoch, over cosmic scales.

Again, the skeptical reader may well ask, "Should I believe such absurdity?" And skeptical readers would be in good intellectual company, because most astronomers asked the same question in the mid-1990s. That is when the evidence first began to convince us of the existence of this strange force that we call dark energy or, in less picturesque fashion, the cosmological constant. Once again, cosmologists were dragged kicking and screaming into an acceptance of a bizarre new component of the universe. It is only by seeing the sheer weight of evidence that most (though not all) cosmologists now accept that we will need to live with dark energy, measure its properties, and thereby tame it. The goal is to measure it, understand it, make predictions from it, and in all ways ensure that it behaves reasonably with all the other domestic forces in our stable of physics.

But in a sense we have been here before, in the company of none other than Newton, Laplace, Einstein, and Lemaître. For the observer, the universe is a curious environment. Its capacity to spring surprises is legendary. What we will learn is that, long before cosmic acceleration was discovered in the real universe, the theoretical physicists—seemingly via precognition—had already derived the equations that accurately describe this uncanny behavior. As we noted earlier, there had been a major problem, known since Newton's time: if gravity is the only force acting on large scales, then gravitational collapse, the astronomical equivalent of Armageddon, might be looming just over the horizon. Attempts to address this fear led to the discovery of the cosmological constant.

■ The Invention of the Cosmological Constant or Dark Energy: 1915

Newton bequeathed to his successors a universe in which time tick-tocks absolutely, independently of observers, and space likewise is absolute and exists independently of the objects embedded in it. Einstein overturned all of this, first with his 1905 paper on special relativity, and then again in November 1915, during the frenzied performance at the Prussian Academy of Science in Berlin where he presented general relativity in lectures written at breakneck speed. Einstein was desperate to fend off competition from the Göttingen mathematician David Hilbert, who was within a hairbreadth of deriving the covariant equations (which we show below) of general relativity.

Take a look at this mysterious field equation that Einstein first unveiled on November 25, 1915:

$$R_{\mu\nu} - \tfrac{1}{2}\, g_{\mu\nu} R = 8\pi T_{\mu\nu}.$$

You can see this equation (or a variant of it) on T-shirts worn on campuses by keen physics students. But what does it signify? The

Greek subscripts indicate that we are dealing with ten related equations here, written in a highly compact form called tensor notation. A simple interpretation is possible.

On the left side we have a mathematical description of how the geometry of space-time (space plus time) is distorted (curved, warped) by the masses (planets, stars, galaxies, black holes . . .) it contains. The right side of the equation is based on energy: it tells us how mass moves around in gravitational fields. There is a nice tension here, the choreography of motion in space-time. To the physicist John Wheeler is attributed the quotation "Matter tells space-time how to curve, and curved space tells matter how to move." Did you notice we used the expression "covariant equations" just now, when saying that Hilbert and Einstein were in a race against time for priority? Covariant means that the equations fully incorporate all forms of motion: inertial, rotational, accelerated, and so forth, and they are written in a form so that they are true for all observers, not just those that are stationary (or moving at a constant velocity) with regard to the universe as a whole.

Newton's laws, Einstein had noted, would not be true—as written—for observers who happened to have unsteady motion. It seemed strange to him that there were preferred frames of motion: good frames in which physics behaved properly and bad frames of motion within which the laws of physics, derived in the laboratory, would not hold. To go back to our train analogy, the laws of Galileo or Newton would be just as valid in a steadily moving train as in a stationary one, but they would fail if the train were in the act of stopping or starting. How would a single observer in a dark box, unaware of what is happening in the rest of the universe, know if the frame she was in were good or bad? Would it not be better to have laws of physics that are correct in all frames? So the complicated form of the equation above was a result of the mathematics that Einstein had to learn in order to write the equations of motion so that they would be good for observers in all frames of reference.

Now in Einstein's field equations, instead of the coordinates of space and time being a sort of fixed-grid reference system to locate objects, we have a dynamic system in which space and time are not independent of each other. The dynamics are both determined by the positions of objects and also they in turn determine the motion of objects and the rippling of the fabric of space-time. Gravity is equivalent to acceleration.

Einstein was not able to find any solutions to the field equations that he had developed. He left that to others. For Einstein, the enormous scientific achievement of general relativity was overshadowed for years by personal problems: illness, impending divorce, and food shortages due to the war. But he had laid the foundations for studying the nature of the universe. To develop the astronomical implications he corresponded with Karl Schwarzschild, an astrophysicist, who as director of the Potsdam Observatory held the most prestigious astronomy position in Germany.

Schwarzschild, while on German military service in Russia, somehow found the time to calculate (using Einstein's equations)

Figure 7.1. Karl Schwarzschild (AIP Emilio Segre Visual Archives, courtesy Martin Schwarzschild)

how space-time would curve outside of a spherical star. If the mass of a star could be compressed into a small enough radius (about three kilometers for our Sun), then the equations all broke down: space-time was ripped apart. The result is what we now term a black hole (the name invented by John Wheeler), and these objects have been discovered throughout the universe, with the most massive ones found in the centers of the largest galaxies.

Schwarzschild had found the solution for Einstein's equations that would hold outside a point mass, a black hole. Unfortunately, while on the Russian front, he contracted the untreatable, blistering skin disease, pemphigus, and died in Potsdam in May 1916. Einstein then had to turn to friends in Leiden, Willem de Sitter and Paul Ehrenfest, for guidance regarding the implications of the general theory for cosmology. The Schwarzschild solution to Einstein's equations is still the bedrock foundation for all calculations of motion in a spherically symmetric gravitational field whose mass is concentrated toward the center.

It was soon found that orbits in the Schwarzschild solution were very slightly different from those that Newton's laws predicted. The differences were so tiny that for all except the fastest moving bodies, they would be immeasurably small. But for one planet, Mercury, a difference was adduced that could be measured and, lo, the anomaly that Le Verrier had found in 1859 was predicted by the Einstein-Schwarzschild solution. Thus, one of the two basic puzzles left by Newton was solved. But did relativity contain a solution for the other one, the threatening dominance of gravity on the largest scales over all other forces? Einstein worked on this during the war with his colleagues in the neutral Netherlands.

By February 1917, Einstein had an idea that seemed so crazy he jokingly asked Ehrenfest for an assurance that there were no asylums for mad people in Leiden. In thinking about the kinds of universe that general relativity could accommodate, Einstein rejected a static, infinite universe full of stars as entirely implausible: gravity would be infinite at every point and the cosmos would collapse under its own weight. But a finite universe sitting gently in space was no better and would collapse to its center.

His third option was a finite universe *without* boundaries, the three-dimensional equivalent of our familiar ability to travel as far as we want on the surface of a spherical planet without reaching an edge. Masses in space caused space to curve, so a finite universe, properly arranged, could curve right back on itself. A beam of light could curve back on itself, while appearing to travel in a straight line. The elementary question "What lies beyond this universe?" has no meaning. Einstein rather liked the elegance of this "closed" universe, but it had an important flaw. The theory required the universe to be expanding or contracting, but not static; there simply were no static solutions to his equations. The year is 1917, and only Vesto Slipher (see fig. 1.4) at the Lowell Observatory in rural Arizona suspects that the galaxies are indeed in motion. Einstein, knowing nothing of this, decided to alter his equations in order to allow a static universe. And it was de Sitter, rather than Einstein, who worked through the cosmological consequences of Einstein's fudge.

Einstein changed the field equations ever so slightly, adding a term to provide a positive force exactly balancing gravity on large scales, but having no noticeable effects on laboratory scales or in the solar system. He gave the name "cosmological constant" to the term that he added, and he used the Greek letter lambda (λ) to signify it. The expression *lambda term* survives in some of the vocabulary of modern cosmology. Here is the new field equation with the new forcing term ($\lambda g_{\mu\nu}$) added to the right-hand side.

$$R_{\mu\nu} - \tfrac{1}{2}\, g_{\mu\nu} R = 8\pi T_{\mu\nu} + \lambda g_{\mu\nu}.$$

A justification is given in the final sentence of Einstein's paper of 1917:

> It is to be emphasized, however, that a positive curvature of space is given by our results, even if the [cosmological constant] is not introduced. That term is necessary only for the purpose of making possible a quasi-static distribution of matter, as required by the fact of the small velocities of the stars.

As we noted already, in his reasoning for adding this lambda term, Einstein was wrong for more than one reason. The fact that Slipher's measurements had shown that the universe of galaxies was not static could not have been known to him. But Einstein's motivation was wrong on other, more subtle grounds as well.

The static model that was made possible by the cosmological constant would be incredibly fragile and unstable! Here's why (yes, this is yet another thought experiment). Consider two lumps of matter, for example galaxies, which are widely separated in empty space. In Einstein's new model universe, with the cosmological constant, there is an exact balance of forces, where the cosmological constant pushing them apart precisely matches gravity pulling them together. Now imagine them being moved slightly further apart: the gravity between them becomes weaker (since it goes inversely with the square of the distance), but the force pushing them apart would become stronger (since this new force increases with separation). In consequence there would then be a net force driving them apart. The further apart they separated from one another, the more this net force would grow until they would be flying apart at great velocity. The same process works in reverse: if the two particles were pushed slightly together then gravity would get stronger, the repulsive force would weaken, and soon the particles would be rushing together at great speed. Putting this succinctly, the Einstein static universe is hugely unstable to tiny perturbations. This instability was hinted at by de Sitter, was recognized by both the Russian mathematician Friedman and by Lemaître (who turned it into a virtue) and was proven formally by the Caltech physicist Richard Tolman in 1934. The instability of the static universe that Einstein had invented is of the same essential nature as the instability of a pencil precariously and precisely balanced on its point. The cosmological constant would cause a fatal global instability that would totally destroy the observable universe. Not good!

Einstein was a capable but not a truly great mathematician. However, it is surprising, given his extraordinary physical insight, that he nevertheless seems to have been unaware of the gross instability of the static model he had invented. Looking back, it seems that Einstein immediately felt rueful about this move because it wrecked the elegant simplicity of the field equations that he had battled so hard to derive. He came to believe that he should have left his equations of relativity as initially written and not have tampered with them by adding the cosmological constant. But he was also somewhat slow to accept the expansion of the universe when Hubble's results became known. Initially at least, he thought that Friedman's paper was erroneous, and he told Lemaître in person that his physics was abominable. Science writers have made too much of a line in George Gamow's autobiography (published 1970) in which he recalls a conversation with Einstein he'd had decades before. Einstein is supposed to have "remarked that the introduction of the cosmological term was the biggest blunder of his life." But Gamow was by then an unreliable witness known for exaggeration. All we have from Einstein himself is in a 1917 letter to de Sitter:

> In any case, one thing stands. The general theory of relativity allows the addition of the term $\lambda g_{\mu\nu}$ in the field equations. One day, our actual knowledge of the composition of the fixed star sky, the apparent motions of the fixed stars, and the positions of spectral lines as a function of distance, will probably have come far enough for us to be able to decide empirically the question of whether or not Λ vanishes. Conviction is a good motive, but a bad judge.

■ The Revival of Dark Energy in the 1970s

After that point the road for the cosmological constant became rather bumpy. Lemaître, as we noted, based his fireworks cosmology on the instability of the primeval atom. He used the lambda term (the cosmological constant) in a paper in 1934 where he identifies it with negative energy in the vacuum. Eddington rather liked this and mentioned it favorably in his *Mathematical Theory of Relativity* (1923). But others thought that it was an unnecessary, extra wheel. Were those doubters half remembering Ptolemy's fudge of the *equant*, which Copernicus had detested so much? The cosmological constant fell into desuetude during the classic period of cosmological investigation that Allan Sandage memorably summarized as "the search for two numbers." Why introduce new physics, by making the universe depend on three numbers, when the old physics might solve the problems quite nicely?

Meanwhile, the standard approach to cosmology was not giving the answers that were commonly expected. Many felt that there were reasons to believe that we lived in a geometrically flat universe. It had a beautiful simplicity. In a flat universe the circumference of a circle is always exactly 3.14159 . . . (π) times the diameter; it is never either more or less than this number. In the simple cosmological model that most were considering, in which there was no cosmological constant, the condition for flatness is identical to the requirement that the mean mass density *is* and *always is* at exactly the critical value. This neat relation was developed in chapter 3 and proved in appendix 1, equation (A17). Then, globally, the expansion is exactly balanced by gravity. The basis for this belief was never very persuasive. It arose more from ideology than from rational thinking or measurement. The painstakingly accumulated data were not being compliant with the hopes of cosmologists, and the concept itself hid troubling aspects.

Once again, there was a fine-tuning problem arising from the critical density. In the standard, flat model of the universe the amount of matter is just right to leave the universe after long times in the coasting phase where it glides to lower and lower velocities at a slower and slower rate. The situation (warning, thought experiment!) is as if a rocket is fired from the surface of the Earth upward with *exactly* the escape velocity. Imagine living as a tiny creature on the rocket and wondering what its fate will be. If the velocity is ever so slightly less than the escape velocity, the rocket will fall back to Earth: the observer would probably have crashed and burned. But if it is fired off with greater than the escape velocity, then the observer would soon be zooming out into space and not worrying about gravity at all. It would take incredibly precise engineering to fire off the rocket at the rate that is just right for coasting. In the cosmological case, the current value of the Hubble constant would have to be at exactly the right value and neither a tiny bit more nor less. That seemed unlikely.

But there was a worse problem. The *evidence* was strongly against this model universe. We showed in chapter 3 that there are two ways to determine the fate of the universe. As we have already explained, if the deceleration parameter q_0 were found to have a value less than ½, then the universe would expand forever, but if it were found to be larger than ½, then the universe would expand for a bit longer, stop expanding when it reached its minimum density, and then turn around and re-collapse (see fig. 5.7). Sandage pursued a direct approach by using distant galaxies to see if q_0 was at the magic value of ½. His method was to see how the Hubble constant varied with distance or time. Others, including the Leiden astronomer Jan Oort, took the alternate approach: they were adding up the amount of matter in the observable universe to see if Ω_m was exactly unity. In the standard model that the astronomers were exploring, that had no cosmological constant, one could show that finding $\Omega_m = 1$ would be identical to finding that $q_0 = ½$ (see appendix equation A17). However, both types of measurement failed by a large margin to give the desired answer.

Through to the 1970s the cosmological calculus amounted to the following. If all the normal matter that could be found was added up, Oort found that it undershot by roughly a factor of thirty for balancing the expansion. If, in addition, the calculation included all of the dark matter for which there was evidence, it *still* came up short by a factor of five. What was wrong with that? The universe would expand forever and that was that.

But wait. There was an additional nagging concern. This factor of five was the ratio at our epoch: *now*. Imagine that we had been sentient and had examined the universe at redshift of $z \sim 2$? Then the standard theory for an expanding open universe tells us that, instead of the total matter being one-fifth of what was needed, an observer at that time would have found that the density was much closer to the critical density than it is now.

For the simple model that we are discussing, the density parameter changes over time like this: $\Omega_m(z) = \Omega_{m,0}\,[(1 + z)/(1 + \Omega_{m,0}\, z)]$, where $\Omega_{m,0}$ is the current value of the density parameter and $\Omega_m(z)$ is its value at an earlier redshift. It gets bigger in the past because the redshift z is bigger and smaller in the future. We see that, independent of the current value, $\Omega_{m,0}$, it will always happen that Ω_m gets closer to unity as we imagine going back into the past and the redshift z becomes large; the right-hand side approaches 1 as the redshift, z, becomes large, whatever the present value of the density parameter.

Suppose that Ostriker, Peebles, and Yahil were correct in 1974 and that $\Omega_{m,0}$ is 0.2 (20 percent) at the present epoch. Then at an earlier epoch, at $z = 4$, the density was 56 percent of the critical value, at $z = 40$ it was 91 percent, and at $z = 400$, 4,000, and 40,000 it was 99.0 percent, 99.9 percent, and 99.99 percent. The fundamental point is clear: over the eons of the past, when the universe was expanding by huge factors, the matter density was very, very close to the critical value, but ever so slightly less.

The arithmetic shows that, if we run the universe backwards in time (more thought experiments here) and imagine going into the past, we will find that the matter density inexorably approaches the critical value at very high redshifts, no matter what value it has today. At our present epoch the density parameter seems to be tumbling to a value much less than unity. In the future it will be and will stay near zero. So is it not rather odd that we happen to live in just the moment of transition? And isn't it strange that in the past the total density seems to have been so close to the critical value but not *at* that magic value. This is a serious problem of fine-tuning. It is a logical or philosophical problem. It offends our aesthetic senses; but from the empirical point of view the density is whatever it is measured to be! Bob Dicke at Princeton and others objected that something must be wrong with our calculated cosmological model if it places us at such a special epoch. But in the standard two-parameter model universe there was no escaping this dilemma.

Ignoring the mathematical details, a description of the standard model in the 1970s would have sounded like this. "After the Big Bang the universe was, geometrically, very close to the magic flat model in which the circumference of a circle is π times the radius, and gravity balances the expansion to great precision. But then, much later, the universe noticed, so to speak, that gravity did not quite balance the expansion, and it became weaker and weaker until the present when it is only about one-fifth of what is needed. In the future gravity will be negligible and the galaxies will simply float away from one another unimpeded. We simply happen to be living at just the moment of transition from one state of the universe when gravity was quite important, to a new state in which it will be totally unimportant." That is what the evidence seemed to indicate, and it was not very satisfactory.

This is where the matter rested for decades. There were monumental efforts to find more previously overlooked matter in the universe or to learn how to live with the open, nearly empty model universe that we seemed to find ourselves in, but very little in the way of grand overview and analysis. The cosmological constant was just one of the many wild cards that cosmologists picked up, played with, and yet put down. Then, in 1974, the heterogeneous and creative team of J. Richard Gott, James E. Gunn, David Schramm, and Beatrice Tinsley published an influential paper. It is entitled "An Unbound Universe" and begins with a wonderful quote from the Roman poet Lucretius (c. 99 BCE–55 BCE):

> Desist from thrusting out reasoning from your mind because of its disconcerting novelty. Weigh it, rather, with a discerning judgement. Then, if it seems to you true, give in. If it is false, gird yourself to oppose it. For the mind wants to discover by reasoning what exists in the infinity of space that lies out there, beyond the ramparts of this world . . . Here, then, is my first point. In all dimensions alike, on this side or that, upward or downward through the universe, there is no end. (from *De Rerum Natura*)

The authors took their cue from Lucretius. After weighing all of the available evidence and looking at the inventory of matter and also at the kinematic evidence (from ages of objects and from the evolution of the Hubble constant), they concluded that the much hoped for result that the actual (observed) density equaled the (theoretical) critical density was extremely unlikely. Next they returned to the neglected cosmological constant, which no one had considered for decades. They took it seriously enough to note the possibility of its significance, concluding agnostically that "it remains possible, at least empirically, that there is a dynamically important cosmological constant." The quantitative evidence was simply too uncertain to make definitive statements at this time. A few years later, the daring and perspicacious Beatrice Tinsley, in a little cited paper to *Nature* entitled "Accelerating Universe Revisited," went further and said that the evidence suggested "the only possible Friedman models of the Universe are those with a positive cosmological constant."

■ New Arguments and New Evidence— Dark Energy Confirmed in the 1990s

By two decades later, in 1995, the situation had changed. Many important observed quantities (such as the Hubble constant and the ages of the oldest stars) were known to higher accuracy. A few bold souls went out on a limb to argue that the accumulating evidence strongly supported the existence of a physically significant cosmological constant, large enough to supply the "missing" energy density and allow the attractive "flat" model to be valid. Vast new fields of astrophysical and cosmological enquiry had opened up in the period 1975–95: neutron stars, black holes, coronal mass ejections, X-ray binary stars, and gamma-ray bursts, as well as planetary science and the exploration of the solar system. The technology of large telescopes (on the ground and in space) and their instrumentation (electronic image detectors, radio astronomy, X-ray astronomy) contributed hugely to widening the physical horizons for the observers, as well as opening up invisible zones of the electromagnetic spectrum that had been denied to earlier generations of theorists. In observational cosmology, the redshifts of the most remote galaxies observed climbed steadily and deep surveys of the extragalactic universe were undertaken, with a view to visualizing the overarching structure. Everyone understood that the extragalactic distance scale had to be tightly nailed down because the many published values of the Hubble constant H_0 varied by a factor of two, which was ridiculous.

Two papers published in 1995 summarized the existing evidence. One is entitled "The Cosmological Constant is Back" by Lawrence Krauss and Michael Turner. The second paper was written by one of us (Ostriker) and Paul Steinhardt, "The Observational Case for a Low-Density Universe with a Non-Zero Cosmological Constant." Both papers stressed timing arguments—making sure that the ages of different components of the universe were not internally inconsistent. Given the value of the Hubble constant and knowing the cosmological model, one could compute the time since the big bang (see fig. 7.2). It was more than a little embarrassing for the cosmologists to note that in the standard picture the oldest stars were older than the universe! But, if there is dark energy pushing the galaxies apart from one another in an accelerating fashion, then the universe will seem (from the Hubble expansion) to be younger than it really is, and this problem is alleviated. In an accelerating universe the Hubble constant gets bigger and bigger at later times. If one did not know that the universe was accelerating and fitted that Hubble constant to one of the standard model decelerating universes, one would think that one was looking at the universe at an early time, when it was flying rapidly apart. So, ignorance of the cosmological constant would lead to a substantial underestimate of the age of the universe.

Decades earlier, when Hubble's preliminary estimates of the expansion rate of the universe were very large, they seemed to imply a ridiculously brief age of the universe (compared to the then known age of the Earth). Lemaître had noted that the problem would be alleviated if a cosmological constant were added to the equations. Similarly, with a positive cosmological constant there is more time for structures to grow from small perturbations to their later observed sizes. Also, there is more space along a line of sight to a distant object and therefore a greater chance that there will be a massive object along that line of sight that could produce multiple images (by gravitational lensing) of the background galaxy.

Every apparent empirical discrepancy was reduced or eliminated when one allowed for a cosmological constant. No new ones seemed to appear. In addition, as we shall see, re-introducing the cosmological constant might allow us to recover the holy grail of the geometrically flat universe. The idea was seductive!

Putting the available observations together, Ostriker and Steinhardt concluded that "a Universe having the critical energy density and a large cosmological constant appears to be favoured." These lines of evidence, while persuasive to a few investigators, left most of the community skeptical. More direct evidence was needed, and that was soon forthcoming from two independent groups who studied type Ia supernovae, the exploding stars that Zwicky had proposed adopting decades earlier as standard candles, with the goal of straightforwardly measuring (or refuting) the acceleration of the universe.

■ Dark Energy Fills the Gap, Allowing the Flat, "Just Right" Universe

Before describing the observational work that was conclusive, it will be helpful (for the mathematically inclined) if we return to the intuitive but quantitative cosmological modeling begun in chapter 3, with the mathematical treatment reserved for appendix 1. There we followed the expansion of a small spherical piece of the universe and were surprised to discover that, without using higher mathematics, we could reproduce the essential results of the Friedman and Lemaître cosmological models. We started with Newton's second law of motion and his law of gravity, which allowed us to compute the motion of a point on the shell of a sphere of radius r and mass M acted on only by gravity. It is straightforward to add a force proportional to the distance between two points that precisely imitates the effects of the cosmological constant (see equation A18). We again reserve the details for appendix 1 and summarize the results here.

We find that again there are two possibilities. The first is just a variant of the standard solution, with the velocity of the surface of the sphere declining gradually, as gravity slows the expansion, the sphere finally stops and re-collapses. The only difference that the cosmological constant has made is that the maximum radius is a little bit bigger than it would otherwise have been. But the second kind of solution (A21) to the modified equation is bizarre. This one corresponds to the previous "open" solutions for which the initial rate of expansion was large enough so that gravity never was able to bring the motion to a stop. Then, ultimately the sphere becomes big enough so that the initially unimportant cosmological constant force (which is steadily increasing, since it is proportional to the radius of the sphere) becomes dominant. Then, the sphere rapidly doubles in size, and doubles again and again and again. Gravity becomes totally unimportant and essentially the universe explodes. The dynamics of the various cosmological models are summarized in figure 7.2 with the "open," low-density universe shown in blue and the dark energy dominated universe (which had an earlier start and thus would be "older") in red. The three models that Sandage had considered—that did not have a cosmological constant—were the orange, "closed model, the green "just right," flat model, and the blue low-density model. The lambda-dominated universe has a greater age than any of the others (for the same measured Hubble constant, or slope) and that is why the timing arguments were so important in discriminating among the models.

This extraordinary runaway that we see at late times in the red curve is what is called "inflationary" behavior and is thought to have occurred at early cosmic times. Most interesting is the zero energy "flat" case where, geometrically, the circumference of a circle is always π times its diameter, and where at early times the total energy including the kinetic energy, the gravitational energy, and the dark energy of our little sphere was precisely zero. This solu-

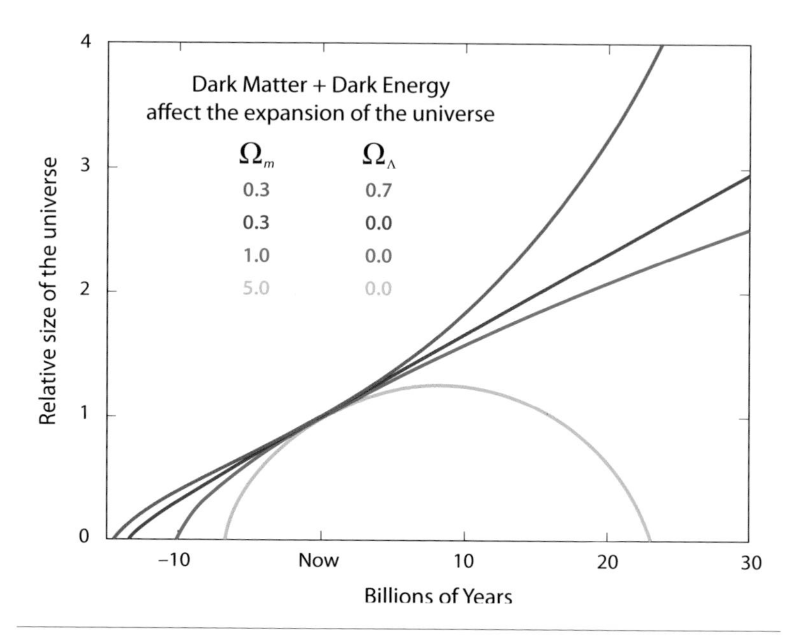

Figure 7.2. Possible scenarios for the expansion (and possibly contraction) of the universe: the bottom orange curve represents a closed, high-density universe which expands for several billion years, then ultimately turns around and collapses under its own weight. The green curve represents a flat, critical density universe in which the expansion rate continually slows down (the curves becomes ever more horizontal). The blue curve shows an open, low-density universe whose expansion is also slowing down, but not as much as the previous two because the pull of gravity is not as strong. The top (red) curve shows a universe in which a large fraction of the matter is in a form dubbed "dark energy" which is causing the expansion of the universe to speed up (accelerate) at late times. There is compelling evidence that our universe is following the red curve. (http://en.citizendium.org/wiki/astrophysics)

tion is both the aesthetically nicest case (having the attractive property that if the universe is exactly flat at one time it always maintains that property) and also, apparently, represents well the universe within which we happen to live. Adopting that case, we show in the last part of appendix 1 that we can define a new parameter $\Omega_\Lambda(t)$ to represent the dark energy as a dimensionless number,

(A23); using that parameter we have the wonderfully simple result for the flat cosmological model, (A25):

$$\Omega_{\text{matter}}(t) + \Omega_{\Lambda}(t) = 1.$$

We see that, if a cosmological constant is allowed, we have a new requirement for a flat universe. The matter density *plus* the energy density, in cosmological units, should always, at all times, add up to exactly unity. While there was empirically no chance that the simpler equation, $\Omega_{\text{matter}} = 1$, could be satisfied, since there was far too little matter in the universe for that (even including all the dark matter), there was a chance that this slightly more complicated equation could be true if there was enough dark energy. Is this attractive and simple equation true? Is it empirically accurate? Is Ω_{Λ} big enough to do the trick? Krauss, Turner, Ostriker, and Steinhardt claimed that the answer was "yes." But their arguments were based on indirect evidence, and more conclusive evidence was soon available.

The most accurate tests have come from the satellite that was the successor to the Cosmic Background Explorer, the Wilkinson Microwave Anisotropy instrument launched by NASA in June 2001 (see fig. 5.6). The effort, directed by Charles Bennett of Johns Hopkins University with scientific leadership by Lyman Page and David Spergel of Princeton, achieved unprecedented precision in measuring the cosmological parameters, with the seven years' summary data giving

$$\Omega_{\text{matter}} + \Omega_{\Lambda} = 1.0023 \pm 0.0055.$$

This is, to date, the most dramatic and precise evidence for the cosmological constant and for the validity of the flat model. One feels that it is almost too good to be true. However, it is not based on the most *direct* measurement.

The simplest direct approach is to examine the changes in the Hubble constant at different epochs using Type Ia supernovae as standard candles, since that produces a direct measurement of the current rate of acceleration. This direct technique uses Zwicky's standard candles in a modern re-incarnation of Sandage's method.

In chapter 3 we described how Hubble set out the overall project and then Sandage, using the Palomar 200-inch telescope (see fig. 3.2) and relatively nearby galaxies, tried to see how the Hubble constant was evolving with time. There were two problems that he encountered: the objects that he used were not bright enough to be seen to great distances, and they turned out not to be truly fixed standard candles. The effort failed. But Type Ia supernovae can be seen to much greater distances than the cosmological markers that Sandage had used, and technology had advanced greatly; telescopes were bigger and the new electronic detectors were now much more sensitive than the photographic plates in use at the earlier time. With these multiple technical advances the attack could be renewed. Observers could use the Type Ia supernovae to learn if there indeed has been acceleration induced by dark energy at late times. Two independent groups have been at the forefront of this research.

One observational effort was led by Carl Pennypacker and Saul Perlmutter; it was based at the Lawrence Berkeley National Laboratory. The other large team was spearheaded and managed by Brian P. Schmidt of the Mount Stromlo Observatory, Australia, with Adam Riess (who worked variously at Harvard, University of California, Berkeley, and Johns Hopkins University). They extended the survey to higher redshifts. It was immediately realized that this method now had the promise to determine Ω_Λ directly, so it was a bit disappointing, when the first results of the Berkeley group were published in 1997, they found $\Omega_\Lambda = 0.06$, which was consistent with the cosmological constant vanishing. The small value that they found made it quite unlikely that dark energy was a major constituent of the universe. However, their sample was small.

But over time the number of measured supernovae increased dramatically, as did the accuracy of the calibrations. In 1998, the high redshift team managed by Riess (who was then at Harvard) astonished the world's press when it announced that their data showed that the expansion of the universe was accelerating. Their team effort was rather large and included Schmidt, Riess, and Harvard astronomer Robert Kirshner, who had been working in this area for some time, and who had been the PhD advisor to both astronomers.

By 2001, both groups had accumulated sufficient high-quality, high redshift data to agree with one another and to convince much of the community that equation $\Omega_{matter} + \Omega_{\Lambda}$ = unity, plausibly described our universe. The two methods, one using the cosmic background radiation and the other, direct measurements of supernovae, now agreed with each other. It looked as if the holy grail had been found; the sum of the matter (mostly dark matter) and the dark energy did total the amount required to produce a geometrically flat universe. Most of the community now agreed that dark energy was the major component of the mass-energy density ($\Omega_{matter} \sim 0.25$ and $\Omega_{\Lambda} \sim 0.75$). Systematic uncertainties, due to the calibration of the Type Ia supernovae depending on redshift, worried some, but the "Lambda CDM" paradigm had been confirmed and was found to be consistent with all of the major cosmological tests. Once again, as with the discovery of dark matter, there were many independent lines of evidence, and they all pointed to the cosmological constant and the dominance of dark energy over all other components of the universe. As we noted earlier, Perlmutter, Riess, and Schmidt were awarded the Nobel Prize in Physics for 2011.

It was a separate issue to ask if this model made physical sense. In 1995, when Ostriker and Steinhardt were assembling the data for their early paper, they visited the Institute for Advanced Study to consult with an old friend, Ed Witten, generally recognized as the guru of the particle physics community. What did Ed think of the cosmological constant? Of course he knew all about the turbulent history of this concept, but he had also given considerable thought to whether a cosmologically interesting value of λ made sense. It did not. The value that the astronomers were contemplating was, from the point of view of basic physics, simply weird.

The constant could be zero, or it could be (but of course wasn't) huge. However, in the natural units made up of the fundamental physical constants (the Planck constant, the velocity of light, and the gravitational constant), the proposed value for λ was ridiculously small. It was hard to imagine the rational calculation that could have given λ the value of 10^{-120}; this is the value that Einstein's cosmological constant must have had to give us the universe that we were apparently living in! If the cosmological constant had been somewhat larger than this, the accelerated expansion would have started earlier and blown the universe apart before galaxies and stars could have been made, and, if it had been much smaller, our tools would not yet have detected it. Why was it "just right" to be interesting astronomically at precisely this epoch, giving Ω_Λ comparable to Ω_{matter}? Here was yet another puzzle of fine-tuning in the cosmos, another embarrassing coincidence.

This dilemma has persisted to the present time, as a conundrum of the most intractable kind, to string theorists in particular and to particle physicists in general. The naïve expectation from quantum physics that Ω_Λ be near unity (rather than 0 or 10^{+120}) has been called the worst theoretical prediction in the history of physics.

Over the last decade there have been many independent and ingenious attempts to provide rational arguments for a cosmological constant of roughly the observed value, but to date not one of them has gained a wide following. We can perhaps find the moral to the story of dark energy in Shakespeare's *Hamlet*. "There are more things in heaven and earth Horatio than are dreamt of in your philosophy." Are cosmologists, like Lewis Carroll's Queen, being asked to believe "as many as six impossible things before breakfast"? We will return to this puzzle in our final chapter, but suffice it to say, the present, standard (LCDM or Lambda Cold Dark Matter) model has passed every test to which it has been put. We seem to have been forced into one of the oddest situations ever encountered in science. We have a model for the universe that really works in the sense that it truly passes every empirical test; yet it is founded on two mysterious, invisible components whose influence is palpable but whose nature is totally obscure to us.

Chapter Eight

The Modern Paradigm and the Limits of Our Knowledge

■ We Have Come a Long Way

What do we know, what do we consider likely, what do we conjecture, and what is it that is frankly unknown to us at the present time? We will now put together the various pieces that have been introduced in the prior chapters, summarizing what has been called the modern paradigm of cosmology, continuing with some of what is exciting but uncertain, and ending, in our final chapter, with what is in some sense most important, the open questions of cosmology. The scientific method, used to construct and test the picture that we have put together, was developed in large part to help us understand the heavens above, and its successes have encouraged us to apply the method to field after field: quantitative observation, the construction of theories to explain and understand these observations, and the testing and refining of these theories with new experiments and observations. From Galileo, through Newton, and then Hubble, to the latest Nobel Prize for the discovery of dark energy, the process has been the same. A proliferation of ideas, with most discarded as false and some very few surviving in the global model.

From the beginning of human consciousness until the twentieth century the heavens had two components: the fixed stars and the wandering planets. Renaissance scholars established order among the planets, discovering and confirming a Sun-centered solar system ruled with admirable precision by Newton's laws of motion and gravity. By the late eighteenth century the fixed stars had been conceptually reorganized into an enormous disk-like assemblage—our Milky Way—and our off-center position in this galaxy was recognized by the early twentieth century. Our situation was found to be ordinary in the extreme. We live on the third planet orbiting an average mass, middle-aged star, which in turn is in a typical orbit accompanying more than a billion stars circling a disk-like galaxy.

The modern cosmological paradigm began to take shape roughly a century ago, with two independent developments. Einstein's general relativity gave us the physical laws that govern the observable universe and taught us that matter and energy must be treated on an equivalent basis. Simultaneously, the astronomical discoveries by Hubble and others proved that we live in an expanding universe, the building blocks of which are galaxies like our own Milky Way, showing us the stage and the actors playing to Einstein's script. For the first half-century the question asked by cosmologists concerned the overall plot. How would the story end—with continued expansion or with a catastrophic re-collapse? But then, as the evidence grew that the whole show began in a fiery explosion, other questions obtruded, begging for answers. How (and why) did the cast—the galaxies—assemble, get their costumes and begin interacting in pairs, groups, and clusters?

The issue of the origin of structure only began to be treated seriously in the 1970s. The effort to understand both the fine-grained detail and also the global dynamics of the universe forced us to the realization that the ordinary chemical elements of our visible world were just a small portion of the matter and energy of the universe. The dominant component that was pulling parts together via gravity was "dark matter" which did not interact with ordinary matter except gravitationally; it existed in and around the normal galaxies. Finally, we discovered that there was yet another mysterious component, "dark energy," that opposed gravity and was found to be pushing galaxies apart from one another with increasing force as time and separations increased. In terms of matter-energy abundance, the dark energy now (A24) overwhelms all of the gravitating matter by a ratio of 3:1; and within the gravitating matter the dark component overwhelms the normal chemical elements by a 5:1 ratio.

The questions came thick and fast. The more clear and specific the model became, the more we realized that questions which we had taken for granted, like the cause of large-scale homogeneity and isotropy, the origin of perturbations and, in fact, the origin of the observed universe itself, were issues that we should address. Every honest treatment of cosmology must end with the still unresolved big questions. But first let us lay out for the reader the detailed map of the immensely successful modern cosmological paradigm whose parts we have assembled in the earlier chapters.

■ The Matter and Energy Content of the Universe

We start with the *stuff* of the universe. Of what is the universe composed? The idea, crystallized in the Roman poet Lucretius' *De Rerum Natura* (*The Nature of the Universe*), is that all of the things we see around us in the natural world can be understood as made up of various combinations of a much smaller number of chemical "elements." Amazingly, this intuitively profound but purely speculative scheme of the ancient world has turned out to be true. By the twenty-first century we are so accustomed to this wonderful simplification of the world that we have ceased to consider it remarkable that the world around us is made up of distinct chemical elements.

Living things are composed of relatively few elements: hydrogen and the abundant moderate mass elements—carbon, nitrogen, and oxygen—with a smattering of heavier elements such as phosphorus and iron. The atmosphere and the oceans are composed primarily of the same few elements as our bodies. Today we know of 92 natural elements and 118 elements in total. Most of the mass of our planet is made of very few elements. Oxygen, a major component of air and water, is almost 50 percent of the mass of the Earth's crust, and the handful of other common elements that we have mentioned add up to more than 99 percent of what we find around us. However, what we find on and in the Earth is not at all typical of the universe, in that the two lightest elements, hydrogen and helium, cosmically the most abundant elements, have largely evaporated into space from our warm planet.

Early in the twentieth century, astronomers looking at the spectral lines seen in the light from the atmosphere of the Sun and other nearby stars realized that these two lightest elements, hydrogen and helium, constitute most of the mass seen in the universe, with the estimate for our galaxy being 97.9 percent. The spectral lines in the light emitted from distant objects allow cosmologists to determine the chemical composition of even the most remote stars, galaxies, and quasars. It is reassuring that no elements have been found in other stars that we do not find on the Sun and the Earth, but the proportions of the different heavy elements vary from place to place. It seems that the highly evolved stars known as white dwarfs may be made mostly of heavy elements, and that fact provided an important clue indicating that perhaps the heavy elements were made in stars well after the big bang.

As we stated in chapter 4, the five lightest elements were undoubtedly made in the big bang itself (see fig 4.3, p. 111). We are confident of that because the detailed isotope ratios of hydrogen, helium, lithium, beryllium, and boron fit so perfectly to the results expected from the 10 minutes of nuclear cooking that immediately followed the big bang. Apart from hydrogen, most of the materials of our bodies, and of our planet, however, we now know were made in stars (starting from the simpler, lower mass, elements), often in the dramatic supernovae events of exploding stars. We summarized in chapter 4 how Herman Alpher first noted in the mid-1940s that the exact ratios of specific heavy isotopes to one another found in ordinary rocks corresponded to nuclear cross-sections. And then, in 1957, Hoyle, Fowler, and the Burbidges showed, in detail, how high-temperature and high-density nuclear cooking within the stars during these postulated (and observed) gigantic explosive events could make the heavy elements.

All of this ordinary matter (the contents of the periodic table) found in stars in our night sky and in the nearby galaxies is called the *baryonic* component of the universe. But are most of the baryons to be found in stars? As late as 1980, astronomers would have answered: "Probably yes." But then in the early 1980s, several groups discovered absorption lines in the spectra of distant quasars that could only have been caused by neutral hydrogen clouds along the line of sight to those quasars. With further study helium lines were also found, and within a decade it was realized that most of space is filled with a cosmic web of filaments primarily made of these elements and containing a smattering of the heavier elements probably ejected from stars by stellar winds and explosions. Figure 8.1 shows the gaseous structures in a typical cube of the universe, with the embedded galaxies, the small dense patches, shown as red dots.

Apparently the efficiency of turning gas into stars is only roughly 10 percent, leaving most of the ordinary, baryonic matter that was made in the Big Bang still floating about in intergalactic space. As we noted in chapter 4, the presently observed ratio of photons in the cosmic background radiation to baryons is just what was expected from the abundance ratios of the light elements. This provides another dramatic confirmation of the big bang model.

The very slight contamination of our universe by heavier elements, with trace amounts of oxygen, nitrogen, iron, and other elements, detected when one looks through the filaments seen above toward more distant stars or quasars, is almost certainly the detritus from the patches of stellar evolution occurring in star clusters embedded in the filaments.

Can we now make an inventory of all the ordinary matter, adding to the observed stars in galaxies of all the gas left over from the Big Bang and not used in the building project? Summing up the mass in cosmic clouds, and every other observable kind of ordi-

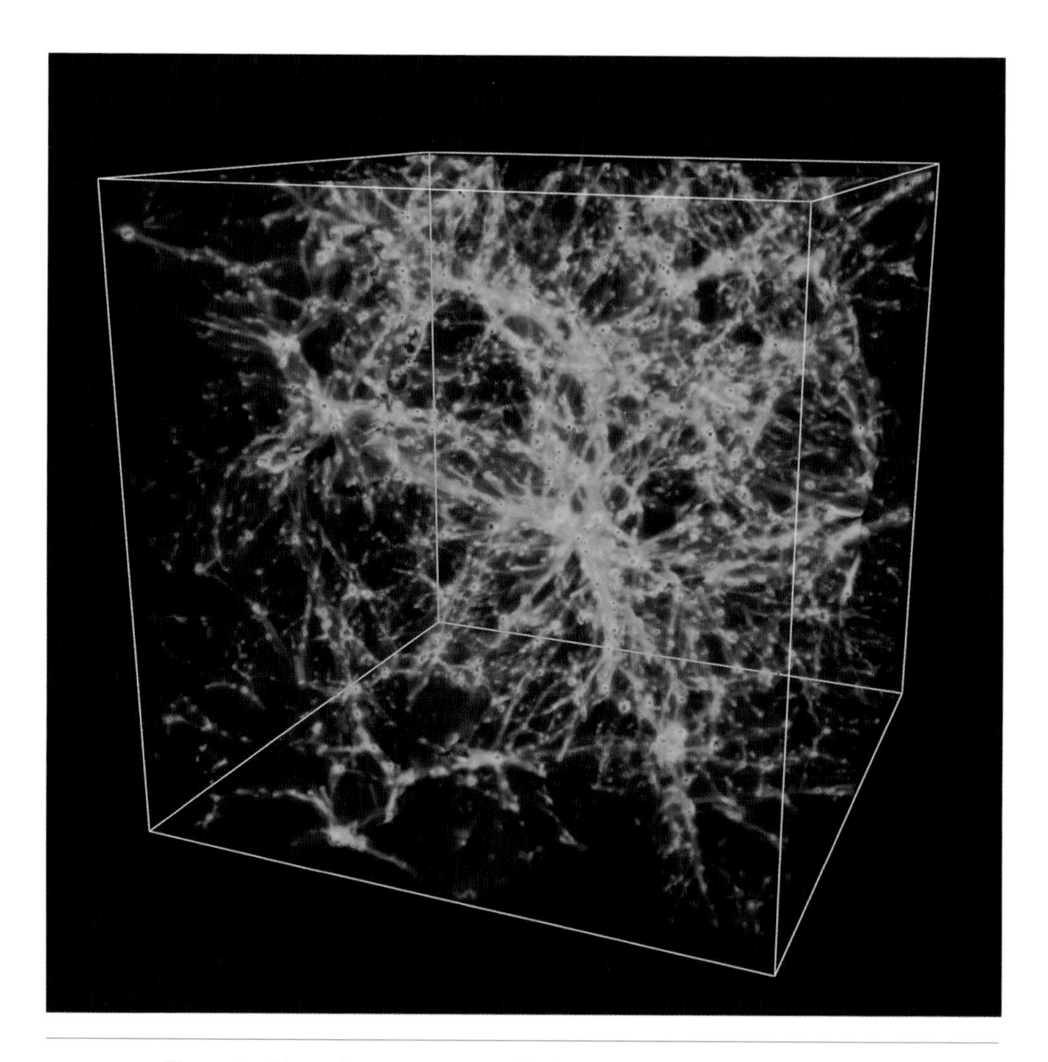

Figure 8.1. Most of cosmic space is filled with a cosmic web of filaments. The filaments of more dense gas surround voids within which stars and galaxies are rare. Where filaments intersect, gas accumulates and galaxies (red) and clusters of galaxies are formed. (Simulation courtesy of Dr. Renyue Cen of Princeton University)

nary, baryonic matter leaves us with a cosmic balance sheet that still falls far short of the level that matters in cosmology, the level of mass density where gravity, emanating from that mass, can slow the global expansion. The ratio between the observed matter and the critical density, which we designate as Ω_{matter}, has the value of about 4.5% (±0.2%, WMAP7). The critical density is the value of the matter density of the universe that would create sufficient gravity to balance the expansion energy, as we showed in chapter 3. Apparently the mass density in the ordinary chemical elements fails by a factor of roughly 20 to slow and reverse the expansion. The familiar chemical elements that constitute our bodies, our planet, and the stars in the sky are apparently not greatly relevant to the grand cosmic inventory of matter and energy.

There is a "good news–bad news" aspect to the results of the baryon inventory. It is very good news that the observed matter density of the universe in baryons is consistent in detail with expectations from big bang nucleosynthesis. But it was treated as bad news that the baryonic content was so greatly below the philosophically neat model universe (described in chapter 3), expected to have a value for Ω_{matter} of 1.000 . . . designated "unity." Recall that the holy grail (for some) was to have a universe in which there was an exact balance of the forces of nature, where gravity just balanced expansion and a circle had a circumference exactly π times its diameter (see the green line in fig. 7.2).

Then, in the 1970s, after the early discoveries of dark matter had been confirmed by many different routes, it was hoped that by adding together the ordinary matter and the more abundant dark matter we could reach a total value for Ω_{matter} of unity. But in fact the most careful accounting could never bring the total to more than about one-quarter of the amount required to achieve the ideal (or desired) mass, with the best current estimate for Ω_{matter} being 27.2 percent (± 1.5%, data from the WMAP satellite). In this period, both of the authors attended more than one colloquium in which scientists plotted on a time chart running through the twentieth century, the ratio of discovered mass to the *required* mass—the mass density to which these scientists were ideologically attached. Since the trend showed such a dramatic increase in the mass of matter already discovered with increasing time, the lecturer would argue that more matter would be found someplace, sometime soon. So we fooled ourselves that we were *almost there.* In fact, there was a period of time (now conveniently forgotten) when many of the sages talked about and even confidently wrote papers in the early 1990s about the "standard model" in which Ω_{matter} really *was* unity.

The ultimately fruitless search for more and more matter did have important by-products. Two quite interesting components of the universe were found, although the mass density in them is not significant on a cosmic scale. Neutrinos, elementary particles postulated in the 1930s by Wolfgang Pauli and Enrico Fermi, and known to be roughly as abundant as the cosmic background photons, were discovered in the 1990s to have mass. They actually contribute a small amount to the cosmic mass density inventory, on the order of 1 percent of the critical density.

We have also mentioned massive black holes. These appear to reside in the centers of all massive galaxies with mass approximately 0.1 percent of the galaxy mass (or roughly two parts in a million of the critical density). We honestly do not know how they were formed but believe that the bulk of the mass in them accumulated from accreted baryons (ordinary matter), and, if this is the case, then they should be included in the baryonic component of the total mass budget of the universe, although of course the baryons can never be extracted from the black holes that devoured them.

The effort to find enough matter to reach the critical value failed. But, as that failure became clear, a totally surprising set of developments revealed that there is still another component of the universe, dark energy, even more abundant than dark matter. Einstein had shown that matter and energy are interchangeable, so the dark energy must be added to the inventory. This powerful component is causing, at the present epoch, an accelerated expansion of the cosmos. The (misnamed) Hubble constant, the rate of expansion of the universe, is not constant but is getting larger rather than declining, as had been expected due to the action of gravity. This bizarre fact has been labeled by many the most surprising discovery of science in the last half century. The best current estimate for the dark energy is $\Omega_\Lambda = 72.8$ percent (±1.6%) and, when added to the previously discovered components, the total matter-energy density really does add up to unity—1.000. . . . In the next section we summarize what this implies for the global cosmology, but it is worthwhile to show in pictorial form (fig. 8.2) the strange history of our discoveries. The orange line segment represents the normal chemical elements, the abundance of which was established by 1950, the black part represents the dark matter, firmly established by the mid-1970s, and the red part to the most recently found and largest part, the dark energy.

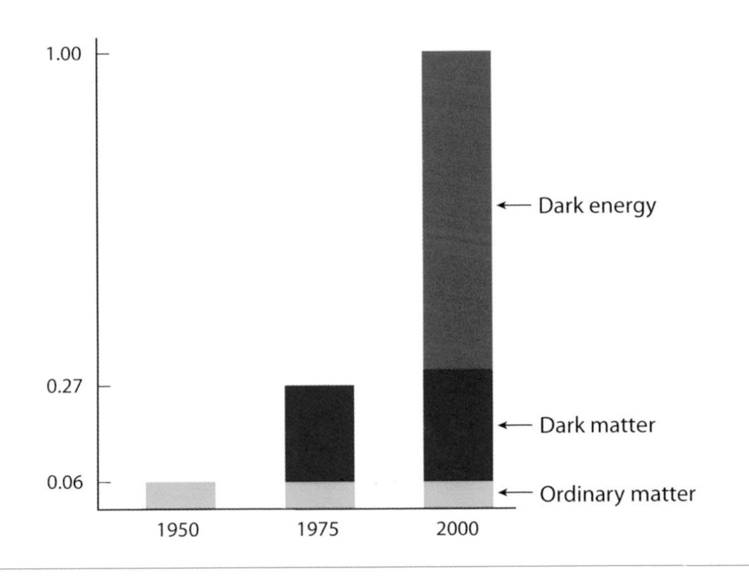

Figure 8.2. The formation of the ordinary chemical elements in the hot big bang and within evolving stars was elucidated in the 1950s–1980s. Dark matter, although proposed in the 1930s, had its existence verified as the dominant material component in the 1970s, and the still larger amount of dark energy, conceived of at the onset of cosmological studies, was observationally confirmed in the 1990s.

Having achieved unity for the total value of omega, including all three now recognized major pieces, it is legitimate to check if there are additional components that could push it over the top. Perhaps there is too much matter and energy in the observable universe as compared to our ideal flat model. There are indeed other small components; these include the black holes, the matter and energy density of the neutrinos that we mentioned in the last subsection, the energy density of the microwave background photons in the background radiation, and indeed starlight which should be included in a very careful audit. The extra amount in these forms is negligible and within the estimated uncertainties of the three major pieces. Other small constituents of the universe will surely be found, but we now really expect the value of unity (for the total) to endure. All the evidence from analysis of the audit confirms that the amount of stuff in the universe is consistent with the value of unity to high accuracy ($\Omega_{total} = 1.002 \pm 0.005$); this quantitative estimate includes all of the small components previously mentioned. We do live in the geometrically flat universe envisioned as ideal by cosmologists beginning with Einstein, and it seems as if this state is enduring. Einstein would have been surprised but probably pleased to discover that the cosmological constant, or dark energy, is the critical ingredient that makes this possible.

The Global Cosmological Solution and the Cosmic Triangle

Before characterizing the structure in the universe, the galaxies, clusters, voids, and so on that fill the sky, let us summarize the global cosmology within which all this structure exists. The consensus among cosmologists is that we do appear to be living in a universe whose large-scale features are as simple as could be: infinite, homogeneous, and isotropic as postulated in the work of Einstein, Lemaître, and Hubble. There is much interesting detailed structure, but no evidence for any asymmetries on the largest observable scales nor for global rotation nor for periodic structure nor for any of the other entertaining variants that ingenious investigators have sought. In that sense, we find ourselves in the "plain vanilla" variant of the cosmos.

Furthermore, when we add up the constituents now, they quite nicely sum to provide the matter-energy density needed for the preferred, geometrically flat case within which the circumference of circles really is π times the diameter. The other neat property of the flat universe is that it stays flat. If the matter and energy density add up to exactly the critical value at this instant in time, then the same was true in the past and will remain true in the future. The ratios of the various components to one another will change over time, but the total will always add up to the magic value, if we are correct that it is now at that value. In any case, that is what the best current data seem to indicate.

Let us now go back to the equation that summarizes the balance of forces in the universe, derived rigorously from Einstein's field equations, after he included the cosmological constant (see chapter 7). If we generalize it slightly to allow for the possibility of curvature, as opposed to flatness, then the equation first derived in 1922 by Alexander Friedman becomes

$$1 = \Omega_{\text{matter}} + \Omega_{\Lambda} + \Omega_{k}.$$

This equation is the dimensionless form of Einstein's equations of general relativity as applied to a universe that is globally smooth (the same, at any given time, from place to place). The equation is rather similar to what we had considered before, and derived in appendix 1 (equation A25), but there is now an extra term, Ω_k, which represents the curvature of space or the net energy density in the universe. General relativity tells us that matter and energy determine the curvature of space and then the curvature of space determines the motion of matter and the flow of energy. If the matter and the energy representing the first two terms in the right-hand side add up to the magic critical value of unity, then the last term, representing the curvature of space, must be zero. We have noted that, observationally, the last term seems to be at or very close to zero empirically, but, whether or not that term vanishes, the total of the three terms must add up to unity or general relativity is wrong.

Now, armed with the good estimates that we have arrived at for the stuff—the matter and energy content of the universe—let us turn back to the classic question of the possible set of global cosmological models that are consistent with cosmic uniformity.

For this purpose, the "Cosmic Triangle" invented by Paul Steinhardt provides a neat and comprehensive way to envisage at once *all* of the homogenous and isotropic world models by their placement in one diagram. It is almost magical that one (admittedly complicated) pyramidal triangle can encapsulate all of our cosmological speculations and show how we homed in on what we believe is the correct solution for the global cosmology. We will be able to locate in one picture (fig. 8.3) all of the possible cosmological solutions that have ever been considered for Einstein's equations and to see exactly where the modern paradigm has settled after a century of investigation.

There is a curious result that we all learned in high school if we studied geometry: that the sum of the three distances from any point in an equilateral triangle to the three sides of that triangle, drawn parallel to the three sides, when added together, is exactly equal to a side of the triangle. Thus we could take a triangle whose three sides represented the three terms in the equation above—the three major components of the universe, matter, energy, and the geometrical curvature of space—and any point within it would be a solution to this equation and represent a specific model of the universe, with the three distances to the three sides representing the three terms in the last equation.

To begin, let us look at the simplest solutions, those for which the cosmological constant vanishes. These are the ones investigated empirically for much of the twentieth century, and they lie along the diagonal line that goes up and to the right from the $\Omega_\Lambda = 0.0$ point found toward the left end of the bottom of the figure; these are the cosmological solutions with no dark energy where the cosmological constant is zero. The half century of Hubble-

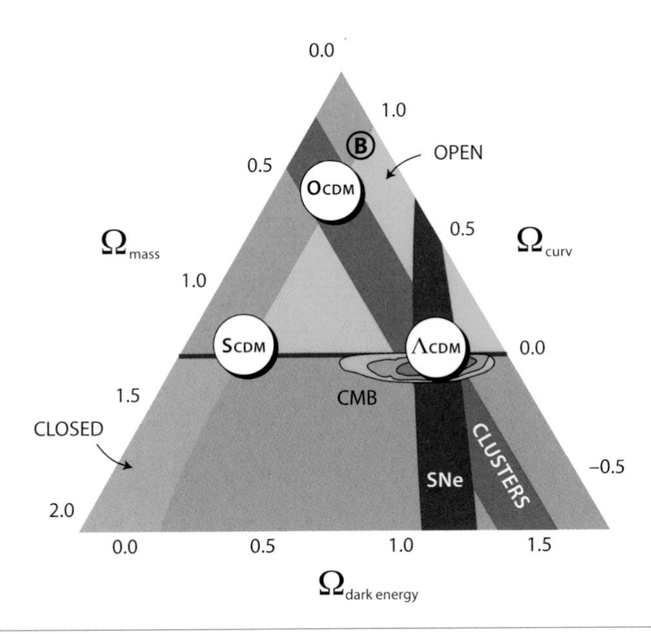

Figure 8.3. The "Cosmic Triangle" which allows one to see and compare all of the standard cosmological solutions. The low-density B ("baryonic"—what you see is what you get) universe without a cosmological constant was the best guess until the 1970s, when the OCDM model containing dark matter was favored. Then, in the 1990s, with the aid of dark energy (or the cosmological constant), it became possible to find the long sought model, ΛCDM, that was geometrically flat ($\Omega_k = 0$) but that fit all observational constraints. (Courtesy of Paul Steinhardt, of Princeton University)

Sandage investigations looked for solutions along this line, because they did not allow for the possibility of the cosmological constant, and, since they only knew about the baryonic component of the universe, the model that they found which best fit the facts was the "lightweight," open universe approximately where the italic "*B*" (for baryonic) has been placed.

This was the model that for decades dominated discussion among astronomers, and it might be called the "what you see is what there is" variant of cosmology. It seems now, in the twenty-first century, quaintly simplistic, and reminiscent of the pre-Copernican model of the solar system in that it essentially assumed that our local universe—the Sun and stars near to us—provided a fair sample of the cosmos. For this model, up in the top corner of the diagram, we see that the matter density of the universe is only a few percent, the universe had positive curvature, and the size of the *B* overlaps both the domains labeled "eventually re-collapses" and "expands forever." The observations that Sandage and others made were not good enough to discriminate between these two possibilities.

Then, after dark matter was discovered and labeled "cold dark matter," the preferred solution drifted down the diagonal line to roughly the place where the "OCDM"—"Open Cold Dark Matter"—circle has been placed. Note that both this region and the baryonic model (*B*) are in parts of the triangle where the universe has a matter density too little to be important, even though it is consistent with Zwicky's observations showing copious amounts of dark matter in clusters of galaxies.

During the 1980s and 1990s, those who, for aesthetic reasons, preferred a flat, zero curvature model (to partially avoid the fine-tuning problem) looked for solutions on the horizontal blue line that goes through the $\Omega_k = 0$ point on the right-hand side of the triangle. These solutions on the horizontal line intersect the $\Omega_\Lambda = 0$ solutions that we have been discussing at the point where we have placed the white circle labeled "SCDM"—"Standard Cold Dark Matter," and that was a happy place to be if you believed (absent evidence) that $\Omega_\Lambda = 0$ and also that $\Omega_k = 0$. For perhaps a decade many cosmologists tried, by every available means, to see if this model could be valid. However, the observational truth was never consistent with this model, and so the community was split for decades between those drawn to one or the other of these two white circles, depending on whether aesthetics or observations most strongly guided them.

We now, with hindsight, know that both communities were holding onto important truths, but that the constituents that they allowed themselves prevented them from coming together. Those—mostly physicists—who realized the seriousness of the fine-tuning problems, insisted that the real universe, whatever the observations seemed to indicate, *must* lie along the horizontal "$\Omega_k = 0$" line labeled "*flat.*" Those—mostly astrophysicists—who knew that all the matter—including all of the dark matter found by Zwicky—was not more than a quarter of the magic, critical number, insisted that the universe lies along the dashed diagonal $\Omega_{\text{matter}} \sim 1/4$ line, which did not pass close to the SCDM circle preferred by the physicists but was in agreement with the observed matter density of the universe.

This quite contentious and unsatisfactory state of affairs persisted until the accumulating evidence, from timing arguments (the requirement that the universe not be younger than the stars within it), supernovae measurements, and many other phenomena convinced even the skeptics that the cosmological constant *must* be revived and treated with respect. The old work by Einstein and Lemaître was dusted off, and solutions that were not on the $\Omega_\Lambda = 0$ line began to be studied again. Very quickly, the neatest possible solution was found. If one proceeded along the horizontal "flat" sequence where Ω_k was identically zero but Ω_Λ was taken as a free parameter to be determined by observation, one came to rest very nicely in the white circle labeled "ΛCDM" for Lambda-Cold-Dark-Matter on the preferred, flat, $\Omega_k = 0$ line.

This model has the matter density that is observed; it is geometrically flat and it predicts a universe consistent with the supernova studies (the blue, nearly vertical, tube) and the timing arguments. It is the modern paradigm of cosmology. All of a sudden it became clear that every observation that had been made was consistent with this model, and in particular, the careful analysis of the cosmic background radiation field by the WMAP satellite (see the red contours labeled CMB) gave empirical justification for it to a higher accuracy than two significant decimal places.

It would be stretching the truth to say that doubt has ceased, and that all students of the problem are now convinced of the accuracy of the flat LCDM model containing dark matter plus dark energy that adds up to the critical value. But today the burden of proof has moved to those who doubt its validity; the consensus is strong that the modern cosmological paradigm is right.

■ In the Beginning

Given this model for the present, how can we plausibly extrapolate it forward and backward in time? Going backwards is easy (see fig. 4.3, p. 110). Earlier in time is the same as higher in redshift. The model is consistent with the growth of structure (roughly redshift six to zero), the dark ages and re-ionization (redshift ten to six), consistent with the cosmic microwave background and the recombination of hydrogen (redshift one thousand), and consistent with the formation of the light elements (at roughly redshift one billion—a few minutes after the Big Bang). Before that, we lose track. It is not that we cannot calculate the evolution of the model before that epoch, but rather, we have no way of testing if our calculations are right or wrong.

However, there is an even earlier epoch where we *must* stop, since before this time we have no idea what the equations are! Einstein's theory of gravity and Bohr's quantum mechanics come into conflict at a very early epoch, and there is a looming KEEP OUT sign over the entrance to that time. This era is called the Planck time (t_P) after Max Planck, the physicist who proposed it, and is defined by the fundamental constants of gravity (G), relativity (c), and quantum theory (h):

$$t_P = \sqrt{(hG/c^5)} = 5.4 \times 10^{-44} \text{ seconds.}$$

Einstein and his collaborators tried to produce a theory of quantum gravity, and more recently Ed Witten and congeries of string theorists have broadened the search for a viable guide to this forbidden land, but as yet there is no consensus as to what is the right approach. We return to this dilemma in the final section.

Extrapolating forward is also easy, but here there are of course no tests whatsoever of our predictions. At our present epoch the universe is just starting to take off into a period of wild exponential growth caused by the dominance of the cosmological constant—inflation it is called, and it is similar to the postulated very early period of inflation—in which all the pieces fly away from one another at an ever faster rate. The stars will all die and the night sky will become cold and empty. Will this happen soon? Not really, but the time scale is not forever. By the time our Sun dies, the final cosmic explosion will have begun. However, to be honest, while the computed evolution back in time from the present through to the early synthesis of the light chemical elements is testable and realistic, the rest of the postulated evolution backward or forward is sheer guesswork and could easily be revised by the next edition of this book.

■ Structure in the Universe

Ignoring our uncertainty over the very distant past and the distant future, we seem to have settled upon a quite satisfactory global model for cosmology—for how space and time are organized and evolve; the fit between the model and the observational facts is almost too good to be true. However, there is still the little matter of the composition of the universe—the actual objects that we find in it. What are they and how did they arise?

By the mid-eighteenth century the English astronomer Thomas Wright and others had realized that the Sun was simply another star like those seen in the disk of stars, the Milky Way, in which the Sun had its place. By the late 1920s, astronomers realized that these stars were organized into a galaxy (the expression nebula was still in common use) like others seen in the sky, and that these galaxies were all moving apart from one another with velocities that were roughly proportional to the distances that separated them. It was also realized that, at the highest organizational level, the universe is uniform and isotropic. Although granular in detail, it is essentially the same in all places and in all directions at a given time in our cosmic neighborhood: looking north, south, east, or west, far out into space, we see roughly the same number of galaxies per unit volume and the same kinds of galaxies.

Of course, when we look really far out into space, to regions billions of light-years away, we are looking back to earlier epochs, and then, while the universe is still the same in different directions, we do notice a distinct falloff in all directions of the observed numbers of bright galaxies and clusters of galaxies. The galaxies were formed, grew and evolved over time. At much earlier times they were smaller, fainter, and there were fewer of them per unit volume at any given brightness. This is apparent to the eye in examining figure 5.3, the Hubble Space Telescope, Ultra Deep Field.

But then, if the universe is roughly uniform, why isn't it totally uniform? Whence comes the structure, and at what epoch did it form? Clearly there must have been some seeds planted very early to have grown into the wonderful variety of creatures that we now see in the cosmic forest called space. At first theoretical astronomers simply took it for granted that the structures (galaxies) existed. The child's question, "Where do galaxies come from?" did not have a place in the intellectual arena. After Hubble had shown how to use Cepheid stars to derive distances to galaxies (chapter 2), they were used, as standard candles, standard meter sticks, beacons along the cosmic road. In the search for the global solution to the overarching questions about the origin of everything, galaxies were not taken seriously as independent objects whose origin required analysis. But the sheer accumulation of information about the galaxies, driven mainly by ever more impressive sky surveys, caused cosmologists slowly to awaken. Awareness grew in the community as it gradually dawned on them that they needed to understand how galaxies had come into being. Otherwise they would be stopped at square one in the attempt to comprehend the history of the cosmos.

Beatrice Tinsley was the first astronomer to begin the modest task of computing the evolution of galaxies based on the evolution of the stars within them. That led them back one further step and the theorists began the serious study of the origin and evolution of galaxies starting from scratch, so to speak. Cosmologists shifted the puzzle of the origin of structure back in time. From what initial state did these building blocks of the universe grow? In 1991, the seeds were found (see figures 5.5 *top* and 5.6). They were discovered, not by chance, but as the result of a worldwide, programmed search into the low-level fluctuations to be seen within the cosmic background radiation. At a redshift of roughly 1000 (roughly 13 billion years ago), very small variations were found in the density of matter and light from place to place. With the help of dark matter, these fluctuations could have grown by gravitational instability into what we see around us today.

But that then raises the question of where did these early, very low-level fluctuations of roughly one part in one hundred thousand come from? The philosophical problem of infinite regress (why, why, why?) rears its ugly head! In cosmological investigations it must always eventuate that the solution to the problem of "How did such and such happen?" will lead back to an earlier problem of the same kind, which, if solved, leads to an even earlier problem. The argument arises if we take the stance that any proposition (for example, "dark matter helps galaxies to form") requires justification (for example, "dark matter arose in the very early universe"). The cosmologist must behave like the little child who asks "Why?" over and over and over again.

What is the correct response? A philosopher might argue that the chain should begin with a belief that is justified a priori, that does not rest on another belief or fact. For a scientist, this looks rather dangerous because it risks moving the search for the *primum mobile* in the direction of metaphysics or the supernatural. Nor can we escape by avoiding the question and simply saying "science isn't done like that." Perhaps common sense is the answer to this deep question. That is to say, we can agree that we know what we know, and we agree that, when progress is slow, we must just keep looking. But we may need to look and to think for a very long time before we get further than we are today. It was centuries before the questions that arose after Newton were addressed and answered: "what, really, is gravity, and in what domain do Newton's laws break down?" The current model is so good—even if it leaves important questions unanswered—and looking every year better and better, that it may be decades before fundamental advances are made.

We may have reached a resting point in our attempt to go back into very early times and determine the precise origin of the cosmic background radiation fluctuations. The general level of consensus tested by observation breaks down, and pure speculation rules in this domain—the Planck era. The most common view is that the tiny variations, required by the fine-grained quantum nature of reality, were amplified and stretched by the gigantic factor of 10^{43} during a period of inflation, which was the early epoch during which fundamental forces caused, for some period of time, an accelerated expansion of the universe in a fashion similar in principle to the current epoch of lambda-dominated acceleration but much more extreme. The observed spectrum of fluctuations is consistent with that overall model. This picture can also explain why, when we look to great distances left or right, up or down, and we see pieces of the universe that cannot now see one another, they look alike. But there are more questions than answers here, and we will come back to this issue and the alternatives in the concluding chapter.

■ The Supercomputer Approach

Suffice it to say that, if we start with the fluctuations actually observed in the microwave background—ignoring the problem of their origin—and put these initial conditions into our computational models along with Newton's laws and the best physical modeling that we can attain, we can test the whole picture for consistency with observation. This process is now a major cottage industry for theoretical astrophysics, with teams of scientists harnessing the world's largest computers to address this problem.

There are three phases in the calculation. At early times, when the fluctuations were small (density variations from place to place were several percent or less), the cosmos looked like the ocean surface on a calm day. Ripples and waves flowed through the matter in a way that is straightforward for us to calculate. But these same ocean waves, when they reach a shallow seashore, where the height of the wave is comparable to the depth of the water, can break. In the same way, gravity, over time (enhanced by dark matter concentrations), caused matter waves in the universe to grow, collide, and break, giving rise to a complicated web of dark matter sheets, filaments, voids, and clusters that we see in figure 8.4. The largest supercomputers working for many months are needed to compute the details that you see in the picture, but the calculations are straightforward and all practitioners agree now on the methods and the general outcome. As we have seen before, several times, unanimity is not a guarantor of accuracy, but in any case we can usually test the results of supercomputer analysis by a comparison with the real world in the sky above us.

A first step is to look only at the dark matter and to fix our attention on the peaks, the highest density regions that are called dark matter halos. Today we have a good understanding of the sizes, shapes, and numbers of these peaks in the matter-density and how they grow and merge over time. These calculations include only gravity as the driving force and so, since dark matter is the dominant mass component, they treat only dark matter and ignore the rest of the normal baryonic (chemical) matter as a component to be added to the recipe later, after the primary structures have been formed.

The idea is that, at the centers of the dark matter peaks, the normal matter collects into galaxies—somehow. Thousands of papers have been written, making detailed comparisons to the observed distributions of galaxies on the sky. In these papers, the authors use approximate methods to estimate how ordinary matter would collect via gravitational forces in the centers of the dark matter halos and form galaxies with the requisite properties. Needless to say, the theorists argue that the correspondence between the computed models and the real world indicates that they are doing this correctly, but the skeptic may be allowed some legitimate doubts. There are so many adjustable parameters in the theories being promulgated that it would be difficult to prove any of them wrong, that is, they are not falsifiable. This problem is especially acute since the free parameters in the "semi-analytic" approach are set by the requirement that the final model fits well to the real world.

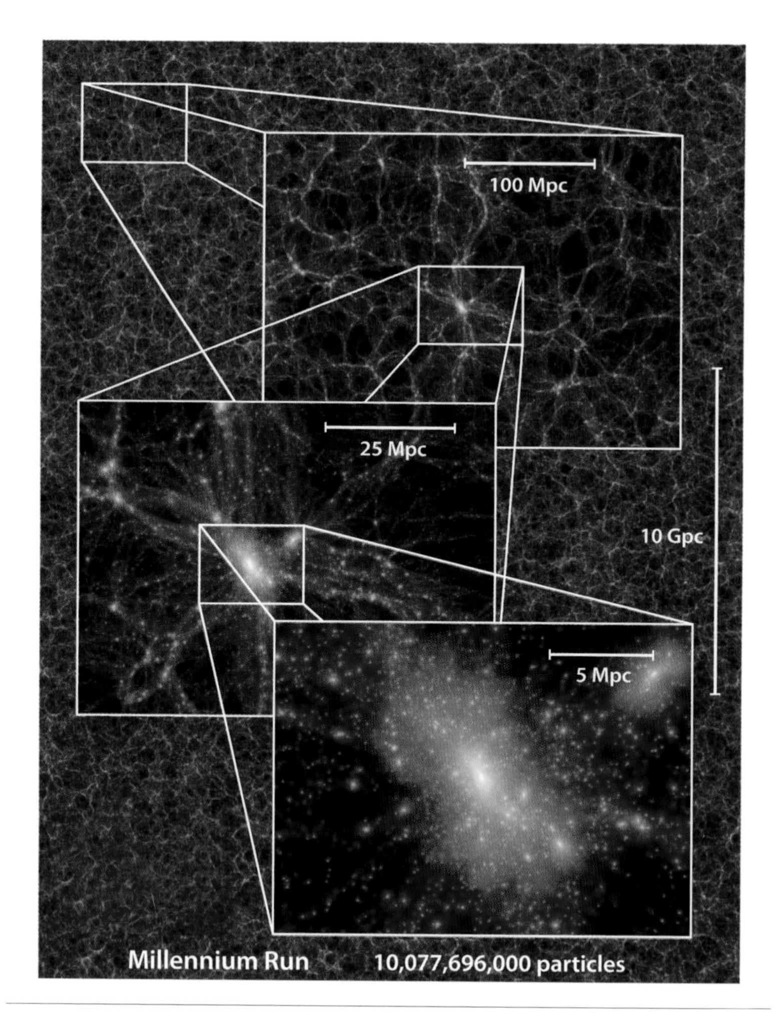

Figure 8.4. Cosmic structure as seen in one of the largest numerical calculations of the universe of dark matter called "The Millennium Simulation." (Courtesy of Dr. Renyue Cen of Princeton University)

There is a more direct but much more difficult approach to the problem. One could put into the calculation all of the baryonic physics, all of the radiation fields, and make hydrodynamic calculations (similar to the ones made by designers of aircraft). One would need to include even the radiation output from the galaxies and the stellar explosions that occur within them and how the forces from these processes compete with gravity in moving gaseous matter around. This proposed set of computations represents a huge job, both computationally and with regard to the physical processes that must be included—it is as big a scientific project as any that has ever been contemplated, and it is under way now at many of our universities and government laboratories.

There is tangible excitement in the enterprise. To play in this game one needs to know, not just the cosmology derived from Einstein's equations, but all of the physical principles and knowledge that have been learned on Earth, from the solid state physics needed to understand cold interstellar grains (at 10 degrees K) to the plasma physics needed to model the very hot gas in clusters of galaxies (at 10^8 K).

The results can be translated into predictions for what optical, infrared, and X-ray telescopes have seen and will see. The Hubble Space Telescope was the first and best known of many instruments—space- and ground-based—that have let us see out into the depths of space and to the earliest epochs of time. The tests of our cosmological picture and our modeling of galaxy formation are manifold. We should be able to predict in detail the local world of galaxies that we see about us, the masses of these systems, the numbers of them, their spatial distribution, their velocities, and even their internal structures.

Not only that, but since we can look far out into space with large ground and orbiting telescopes, we can see far enough back into time to view—directly—the universe as it was when it was half or even less than that fraction of its current age. So, we should be able to directly observe the formation and evolution of galaxies over cosmic time and to compare these with our computer simulations. How well is it going? Of course the answer is far from perfect. With every problem that seems to be solved, we inevitably discover several more critical issues that remain unsolved. But the numerical simulations do seem to be able to predict the formation of galaxies with the right masses at the right epochs at roughly the right abundances. And it is getting better and better. The latest, high-resolution, hydrodynamic simulations seem to predict the observations of the intergalactic medium (fig. 8.1) and even the detailed interior structure of massive galaxies (fig. 5.8), with moderate accuracy. We are doing something right!

We will end this outline of our successes with one more triumph of the standard model. We saw that the "spectrum of fluctuations" of the microwave background seen at high redshift by satellite measurements had in it bumps and wiggles at various definite wavelengths determined by the size of the universe when electrons and protons recombined into hydrogen atoms, the universe became transparent, and the matter was released from the iron grip of the intense radiation field at high redshift. Well, these very low amplitude fluctuations (with peak to trough variation of one part in a hundred thousand) must have grown over time, due to the classic gravitational instabilities that we have described; and they should now be visible in the positions of galaxies in the heavens. Are they seen in the sky? Two of the optical surveys that we described, one called the Sloan Digital Sky Survey (SDSS), done in New Mexico and one called the 2dF Galaxy Redshift Survey (2dfGRS), conducted at the Anglo-Australian Observatory, studied the positions and velocities of tens of thousands of bright galaxies in the northern sky, and dramatically confirmed the prediction in two papers published simultaneously early in 2005. These experiments to find the "baryon acoustic oscillations" were such a success that they will soon be repeated by large international consortia of scientists working with far better instruments on much larger scales to much higher precision.

The Lambda-Cold-Dark-Matter flat model of the universe works. It works really well. Every prediction that has followed from it, when tested, has come out true. But there are fundamental questions that remain unanswered, and we turn in this last chapter to some of those big unanswered questions, those left for future generations of students to decode.

Chapter Nine

The Frontier: Major Mysteries That Remain

■ Dark Matter

By the normal standards of science, we have achieved an astonishing level of success in our search for a viable cosmology. We have an elegant new model that can address every well-defined question. When we build a new instrument to study some aspect of the extragalactic world—for example the large-scale distribution of galaxies—we can predict what it will see; and our predictions turn out to be correct. The uncertainties in our model steadily decline as more and more refined measurements are made, and it now seems as if the simplest, "vanilla flavored," version of the LCDM paradigm is best. Will the situation remain that way, as it did for hundreds of years after Newton's work? Will it be centuries before we (inevitably) learn that there are weaknesses in our magnificent structure indicating that what we have understood is embedded in a bigger truth? In fact, even at present, an honest look at our current model shows that we are profoundly ignorant about the basic underpinnings of the modern paradigm. What, after all, is dark matter, and what is the even stranger dark energy? And what is the origin of the fluctuations that seeded the growth of structure? The short answer to all of these is: We do not know. Let us look in this final chapter at some of the current attempts to answer these questions.

First, it would be very nice if someone produced a bucket full of dark matter and said: Here it is, take a look at it and do some experiments on it. That is not about to happen. But there are numerous very careful searches under way to see if the dark matter, which we believe pervades our galaxy and so must pass through our own planet, can be detected in the laboratory. These laboratories are placed underground, in the deepest mines, to screen out signals from cosmic rays, human interference, and anything that might give a false signal.

Dark matter must interact so weakly with ordinary matter that it can penetrate miles under the surface of the Earth and hopefully register there (if only very weakly) on ultra-sensitive detectors. The idea is that, if and when a dark matter particle passing through the overlying earth chances to hit an atom in the detector, the recoil from the impact will be recorded as an event. And nothing *but* dark matter can penetrate thorough so much rock. A popular candidate for dark matter is the weakly interacting massive particle, the WIMP, which has a good pedigree in particle physics and might easily be produced at the right time in the right amounts. A recent tabulation by Richard Gaitskill listed 27 completed direct detection experiments, 21 ongoing experiments, and 17 more in the planning stage. Sensitivities of these experiments to the many different proposed types of dark matter are improving rapidly, and several ongoing efforts are so sensitive that a detection might reasonably be expected, if our understanding of particle physics is roughly correct.

What are the results? There is only one claim of the likely detection of dark matter. The DAMA/LIBRA collaboration has a target of 250kg of extremely pure sodium iodide, wired with detectors embedded in the Gran Sasso National Laboratory sitting under 1,400 meters of dense granite rock in central Italy. The collisions between incoming dark matter particles and the crystal detectors should generate light signals that are picked up by devices called scintillators. These apparent detections have varied over the course of the years, showing exactly the modulation that would be expected from the "rain in the face" effect. Suppose that in a driving rainstorm one walked around in a circle, taking several minutes to complete a circuit, recording minute-by-minute all the raindrops hitting one's face. When walking into the wind, the rain would be maximal and when walking with the wind the rain would be minimal.

In the same way, the solar system is traveling around the galaxy in a circular orbit and making its way through a headwind of dark matter particles. Then, as the Earth goes around the Sun, it is sometimes heading directly into this wind and then six months later it is heading away from it, so a dark matter signal should show just the annual variation found in the Gran Sasso experiment. The first reported detection was in 1996, of a 1.6 percent annual variation that was quite statistically significant. An upgraded version of the experiment continues to find a highly significant result that is consistent with dark matter detection. While it is puzzling that none of the other experiments have detected such a strong signal, there are possible interpretations that a particle of a specific mass (roughly eight proton masses) and cross-section ($\sim 2 \times 10^{-40}$ square centimeters or fifteen orders of magnitude smaller than the classical cross-section of an electron) could account for the positive result and yet not have triggered other experiments. This claim will surely be tested, so stay tuned.

■ Dark Energy

What about the dark energy? If it is really just Einstein's star-crossed and controversial cosmological constant, we will just have to learn to love it and live with it. No proposed tests that we know of can provide independent evidence for this form of dark energy, other than the cosmological force itself, and we will simply have to wait for some ingenious theoretical physicist to make the bitter brew (Einstein did not like it either) more palatable. But there are other proposals, for a genuine physical force field, often called quintessence (in a nod to a fifth element proposed by ancient philosophers) that could leave other traces. As noted earlier, the cosmological constant has a very counterintuitive value when expressed in units of the fundamental physical constants of general relativity and quantum mechanics: it has a weird and quite tiny value. Any respectable theoretical physicist will tell you that it should either have the value of exactly zero or otherwise be much, much bigger than it is—120 orders of magnitude bigger to be specific. The observationally determined numerical value of the cosmological constant seems quite unreasonable to particle physicists. Thus it is attractive to hope that the dark energy may turn out to be some variety of quintessence rather than Einstein's cosmological constant. The hope is that, if the dark energy is due to quintessence, then the particle physicists can come up with a reasonable explanation for why the force has the value that it does. The evidence to date is not reassuring, but major new observational programs are being proposed that could, in principle, distinguish between the two possibilities.

There are even theories that couple quintessence, the force field (not the constant), to the dark matter force field, so that the number of cosmic mysteries at least might not be increased by dark energy, if this type of quintessence were the cause of the current cosmic acceleration. In other words, the same mechanism might somehow produce both dark matter and dark energy. If a clever physical mechanism were discovered to relate the dark matter and dark energy fields to one another, this would reduce, also, the amount and nature of the required fine-tuning in the universe. With these incentives, the theoreticians have been busy. We are aware of more than fifty papers published in scientific journals in the year 2010 proposing variants of quintessence and showing how it could be consistent with all that we know already. There are far too many to summarize here, but it would be fair to say that, while no one of them has gained a large following or produced telling evidence in favor of its particular flavor of quintessence, the field is lively and promising.

It seems to be generically the case that, if one takes as a given the overall cosmological tests produced by the Wilkinson Microwave Anisotropy Probe and baryon acoustic oscillations (which look at very large-scale waves in the universe), experiments and other global constraints, then the plausible quintessence models predict more structure, such as bigger clusters of galaxies, and more clusters of clusters, than do models with only the cosmological constant. Thus, in principle, with much more work, these theories are testable and falsifiable. We may, in a decade, be able to replace the cosmological constant with something more plausible to explain the dark energy. If Einstein were around to see that happen, we can imagine that he would be pleased rather than mortified by the demise of his cosmological constant.

■ Inflation

Next let us say a few words about inflation, the idea that a moment after the Planck, quantum epoch of cosmology (10^{-43} seconds), the universe had a bout of runaway expansion that left us with a uniform, isotropic, and geometrically flat universe having just the spectrum of fluctuations that we now observe in the microwave background sky.

That picture, first proposed by Alan Guth in 1981, is incredibly attractive to cosmologists because it seems to solve a host of rather deep problems. The standard big bang model that we have been describing works amazingly well so long as it is started off in the right way. In the early work of Einstein, Lemaître, and others, it was taken by fiat that the most obvious starting conditions would be that the universe began as homogeneous and isotropic. But why? And why would parts far apart from one another know how to be the same? If the time required by light to travel between the two parts is greater than the age of the universe, then no signal from one region could have reached the other region and any similarities between them would be coincidental. We gave (in chapter 5) the simple metaphor of pouring water into a glass and noting that the level becomes flat; but that takes time, the time for the water on one side of the glass to sense the level of the water on the other side of the glass and to reach a state of equilibrium.

Whatever the logical problems, it does appear that in the universe it happened that parts that were never in causal contact (in the Hubble-Sandage cosmological models) were similar in nature. In these earlier cosmological models, the cosmos magically started out uniform, as if the water level magically started out level. Inflation provides an alternative mechanism to physically ensure that this state is reached. Inflation also gets the spectrum of perturbations right, and right in very great detail. It predicts correctly the relative number of long wavelength and short wavelength waves. It also predicts correctly the relative number of deep fluctuations and shallow ones. And, roughly speaking, it seems to predict the overall level of the perturbations.

Finally, we found that the universe really does seem to be geometrically flat, with the total value of the matter-energy density equal to the critical value to several decimal places. Inflationary models predicted all of that. These are great successes, but further examination is troubling, as many theorists have pointed out, and the nice metaphor of the glass of water is a little misleading since the inflating universe turns out to be quite unstable.

All of the criticisms of inflation boil down to the same point. Inflation can do all of these good things, but it does not *necessarily* do them, and in fact, for the specific versions of the theory that have been studied, it typically does a lot of bad things as well. The problem is that, once inflation starts, it continues forever. The only places where inflation stops are in islands of space-time, whose number grows to infinity as inflation continues. There will be an infinite number of "good" islands in the overall universe that do look like our universe, but there will also be an infinite number of "bad" islands, where the conditions are radically different from what we observe. The regions that keep inflating are empty and uninhabitable, and most of the volume of the overall universe consists of bad islands. Once again, it takes incredible fine-tuning of inflation to give us the world that we live in. The recurring Goldilocks problem reasserts itself. Whether these criticisms show a fundamental flaw in the inflationary paradigm, a flaw from which all inflationary models would suffer, is something that we do not know. It may be that the right inflationary paradigm is about to be invented by a clever young graduate student as these words are being written.

The problems of the standard inflationary picture led two physicists, Paul Steinhardt and Neil Turok, to invent an ingenious alternative in 2002, a cyclic model universe that may explain the homogeneity and flatness of the universe by a different mechanism and thereby avoid the apparent defects of the inflationary paradigm. In this cyclic picture, the Big Bang is not the beginning of space and time but, rather, a moment when the universe switches from contraction to expansion and a tremendous amount of hot matter is created. The smoothing and flattening occur during a long period of slow contraction before the Big Bang so that the universe already has the right conditions immediately after the bang. An especially picturesque version of the theory is one inspired by string theory, in which our three-dimensional world may be a membrane (or "brane") embedded in a space with a fourth spatial dimension and that another three-dimensional world lies a short distance away along the fourth dimension. In this picture, the Big Bang is a collision between the branes that repeats again and again and again at regular intervals of a trillion years or so.

Each collision is a Big Bang, wiping out the previous universe, and creating matter, energy, and the perturbations that induce large-scale structure in a new expanding universe. This goes through the normal phases but ends (for those living on the brane) with a dark-energy-fueled phase similar to what we are now about to undergo, while the branes are at maximum separation from one another. The dark energy is an essential component needed to ensure that the universe is smooth enough when the branes crash again. But then the attractive forces between them again dominate, they re-collapse, crash and burn, and the cycle starts again. During the infall stage there is ample time for knowledge to propagate from one part of the brane to the other, so the new world is born with flatness, homogeneity, and isotropy already in place.

There is a huge amount of work that must be done to see if this attractive picture can stand up to scrutiny. Many criticisms have been leveled at the picture and many responses to these criticisms have been formulated. The jury is still out, but the cyclic model remains a very interesting alternative to the inflationary paradigm that may still be viable. In a sense this is a modern, quantitative upgrade of the earlier picturesque but rather naïve cosmology in which a big bang is followed by a big collapse and a big bounce back, or of the still earlier Hindu cyclic universes that stretched over the eons.

■ Giant Black Holes

Often weighing more than a billion solar masses, we find them lurking at the centers of almost all massive galaxies. They seem to know their hosts and their hosts know them, or else the consistently tight proportional relation between the masses of the black holes and the stellar components (roughly 1:1000) would be difficult to understand. Most of the time they lie quietly—in fact they are so low in luminosity almost all of the time that it is difficult to understand it. But sometimes, like the proverbial sleeping dogs, they wake, and erupt wildly, emitting radiation in all wavelength bands, powerful winds, and jets of relativistic particles, with the total output in bursts being very roughly 10^{40} times greater than a volcanic explosion. When they are in the eruptive state we call them *quasars* or active galactic nuclei, and there is no good understanding yet of what makes them explode. Clearly the presence of fuel, that is to say ambient gas, around them is necessary, but opinions differ on what provides that fuel. It could be mergers of galaxies, or various instabilities in the surrounding gas that feed the frenzy. The seeds for these monsters could be the several solar mass black holes that are made in the explosion/implosion events that end the lives of massive stars, or it could be that some are primeval, made before stars and galaxies were formed by instabilities in the universe, or, or, or. . . . We simply do not know. And we also do not know if this is a big mystery, tied to cosmology as a whole, or a relatively smaller mystery on the scale of the life cycle of stars.

■ Fine-Tuning

While we are confessing to confusion and weakness of thought and spirit, let us remind ourselves and our readers of one repeated problem that has come up again and again in our account: cosmic coincidences or fine-tuning.

In a popular account, *Just Six Numbers*, Martin Rees has written memorably of the dimensionless numbers in cosmology. One of his six numbers is the ratio of the number of atoms to the number of photons per unit volume of the universe. What is it that has set such a number to have the value that it has? It could have been much bigger or much smaller. Why is it that the cosmic fluctuations in the microwave background are roughly one part in 100,000 and not one part in 1,000 or one part in 1,000,000. If they had been smaller than they were, galaxies would not yet have formed; if much larger, collapses might have proceeded to massive black holes at early epochs. Why is the ratio of dark matter to ordinary matter roughly 5:1? It could have been 5,000,000:1 or 1:5,000,000. Actually, there are more than just the six numbers that Rees has discussed which involve uncanny coincidences, and which if much larger or much smaller in value would have resulted in universes vastly different from our own, universes within which the evolution of life as we know it would be incredibly improbable.

Currently we have no idea what has given the parameters of universal physical law the values that they have. That leads to a litany of unanswered questions. Did physical laws establish these numbers, but we do not yet know what those laws are? Or was it the anthropic principle, which we described earlier in a disparaging fashion as interesting but not scientific, that accounts for the values of these apparent constants of nature? Are these numbers what they are because, if they were different from what they are, we would not be here to ask about their origin? Or should we go back to the wisdom of the eighteenth-century Enlightenment philosophers, including our own sage Benjamin Franklin, and depend (perhaps ironically) on a benevolent and omnipotent deity to have set the stage for the universe? Or is it just chance? It is not a *uni*-verse we live in but a *multi*-verse, containing an infinite number of unconnected space-time regions. The questions are endless. But by now we have clearly left the domain of science and have drifted into the realms of philosophy.

■ Summing Up

Let us come back to present time, present knowledge, and an attempt to sum up and say what we know and what we do not know about the cosmos in which we presently live. Have we succeeded or failed? By the normal standards of science, the success is stunning. Living on this little speck of dust that we call planet Earth, we have surveyed the universe around us to ever greater depths, discovering other planets, stars, black holes, galaxies, clusters of galaxies, and the patterns on the sky which show the seeds from which all subsequent structure developed. We have constructed a quantitative theory for the universe that passes every test to which it has been put. It can be tested and found wanting (falsified) on many levels, but it has not been falsified, yet. Every time there is a new, more precise observation or a new calculation of expected properties, we find better and better congruence between expectation and the real world. This is a pretty impressive list of successes.

But there is a fundamental level at which our work is profoundly incomplete. We understand the minor components of the universe, the ordinary chemical elements and the fields of photons. But the dominant components—the dark matter and dark energy—remain totally mysterious to us. This is the heart of darkness at the center of our understanding of the cosmos. The origins of both of the major components of the universe and the seed fluctuations that grew to become the things we see about us continue to baffle us. The eye of the beholder can thus judge success or failure, but it is surely obvious to all that the undertaking has succeeded beyond the wildest dreams of the early scientists who led the way, while leaving ample room for the discoveries by future generations of cosmologists.

Appendix One

Cosmology without Relativity

■ In a Smooth Universe, Understanding One Small Piece Tells the Whole Story

We are now going to focus on a single sphere of radius r that is tiny compared to the observable universe and is expanding with the universe in its neighborhood. Whatever is the fate of that little bit should be the same for all the other little bits examined at the same time and thus for the universe as a whole. By the way, we use the adjective *tiny* to mean very small compared to the entire universe, but of course it could be very large compared to the Milky Way galaxy and contain a volume big enough so that it can be considered to be an "average" piece of the local universe. It would be bigger than any piece of the universe that Hubble ever studied and of the order of 300 million light-years in radius. Cosmology requires us to think big.

How does our little test sphere behave in the expanding universe? We will imitate Einstein here and perform a little thought experiment. The first point that we note is something proved by Newton—that there is something magically simple about spheres. If we imagine cutting out of the infinite, uniform universe an empty sphere with an ice cream scoop and standing inside the hol-

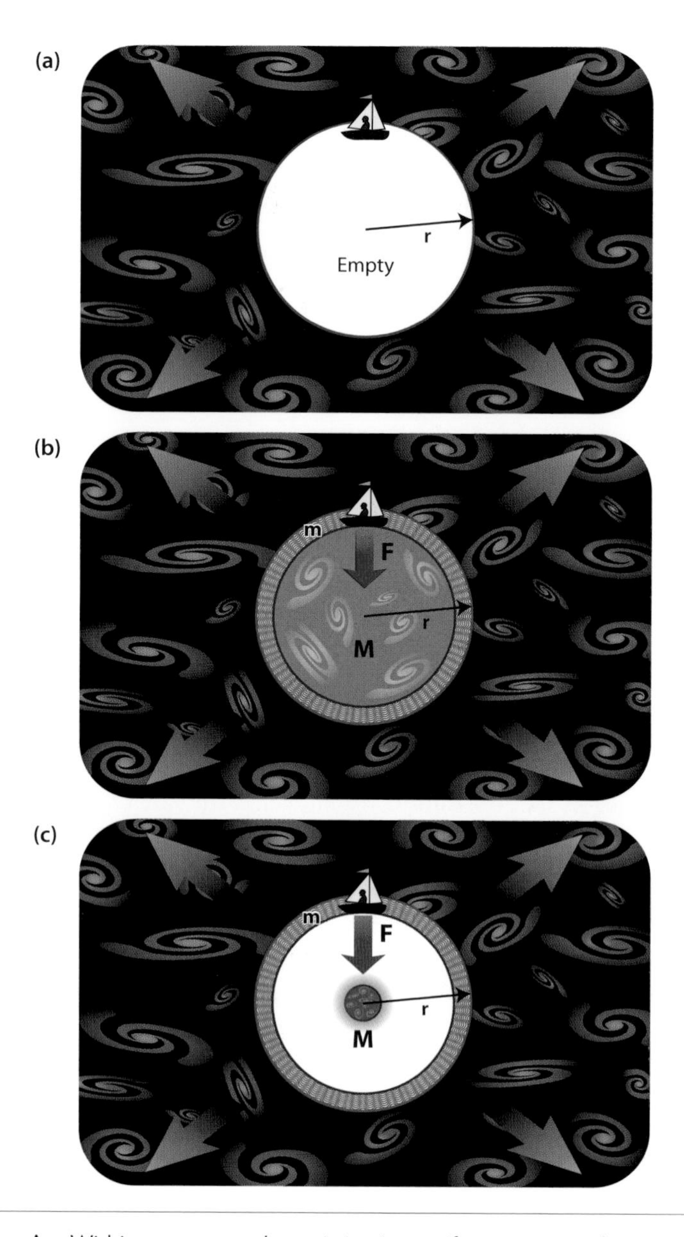

Figure A.1. Within an empty sphere sitting in a uniform universe there would be no gravitational forces (a). If the sphere is filled (b) the gravitational force pulling toward the center is the same as it would be if almost all the matter were concentrated to the center (c).

low space, which way would the force of gravity pull us? The answer is obvious (but not so easy to prove). There is no gravitational force at all in any direction. You can understand this intuitively, since there is no preferred direction and hence there is no direction in which a supposed gravitational force could be pulling.

Now imagine putting a point of mass M at the center of the empty little sphere. Clearly, if we lived in this almost empty region we would now feel a force pulling us toward the center. Newton's law of gravity tells us that the force would be proportional to Newton's constant G, to the mass M, and inversely proportional to the square of the distance that separated us from the center of the spherical region. Now make the point bigger, and make it instead a small sphere at the center. Nothing changes, since the mass stays the same, and Newton showed that, gravitationally, a sphere acts, for those living outside it, exactly the same as would a point mass at the center of that sphere. Let us now expand the interior sphere until it almost fills all of the volume and still has mass M, and let us put a very thin shell of matter at its surface of the same density but having mass m that is very small compared to M ($m << M$).

If we lived on this shell we would still feel exactly the same gravitational force, $F = GMm/r^2$, where we have included in the force the factor m for the mass of the little shell on which we live. We have arrived at the picture, step by step, that is shown in figure A.1. We can think of the shell as the oceans (of mass m) covering the Earth (of mass M) and the observer living in a small boat on the ocean. But this metaphor, while correctly describing the geometry, may give the wrong impression since the spherical piece of the universe is expanding, more like an inflating balloon than like the solid earth.

Let us look at the motion of the sphere, and its outer shell, where we have placed the observer. We start with Hubble's law, assuming that the sphere is expanding with the rest of the universe. The velocity of expansion of the little sphere is governed by the cosmic rule that Hubble discovered, since we want this to be a representative piece of the universe. We multiply the radius r by the Hubble expansion constant H in the same way that the velocity difference between two galaxies is proportional to Hubble's constant and we have Hubble's law:

$$v = Hr. \tag{A1}$$

The fate of the ball thrown upward depends on the energy imparted to the ball, so let us now ask: what is the kinetic energy of that shell due to its motion? From the definition of kinetic energy, E_k, one-half × mass × velocity squared:

$$E_k = (1/2)\, mv^2 = (1/2)\, mH^2r^2. \tag{A2}$$

This is a positive number and so the shell would expand steadily (by participating in the expansion of the universe) were it not for the force of gravity pulling it back due to the mass of matter interior to it. To see how effective gravity will be in retarding the expansive motion, we need its force. This is given by Newton's law of gravitation: $F = -\,GmM/r^2$, which gives for the gravitational energy of the shell, E_g,

$$E_g = Fr = -\,GmM/r = -\,(4/3)\,\pi G\rho mr^2, \tag{A3}$$

where we have used the usual relation that the mass, M, is the product of the density, ρ, times the volume of the sphere, $(4/3)\pi r^3$. The gravitational energy is negative, since you have to put energy into moving something upward in a gravitational field.

How will the velocity of the expanding shell change with time? If we answer that question, we are beginning to do cosmology. We start by calculating its total energy E_{total}, adding together the kinetic energy and gravitational energies (from equations A2 and A3):

$$\begin{aligned} E_{\text{total}} &= E_k + E_g = (1/2)\, mv^2 - GmM/r \\ &= (1/2)\, mH^2r^2 - (4/3)\,\pi G\rho mr^2 \qquad \text{(A4)} \\ &= (1/2)\, mr^2\, [H^2 - (8\pi/3)G\rho]. \end{aligned}$$

Newton showed that total energy will be "conserved"; it will stay constant as the motion proceeds. Also, as we noted, he also showed that the matter outside the sphere (which is also expanding with the rest of the universe) will have no effect on the motion of the little sphere. Now we can ask: will the sphere expand forever or will it expand for a while, reach a maximum size and then re-collapse? We see that, if the total energy is at some time positive, it will remain positive, and that means that the velocity cannot go to zero, which is what would happen if the shell were to reach a maximum size, stop, and then reverse its motion and collapse with negative velocities. Thus, if the energy of the little shell given by equation (A4) is positive, the shell will continue to expand. Looking at the second version of (A4) we see that this happens if H (the expansion rate) is big enough or ρ (the average density) is small enough. But, if the balance is opposite to this, and the total energy given by equation (A4) is negative, then gravity wins, and the shell must ultimately stop expanding and re-collapse. So the criterion for it to be (just) able to expand forever is simply that E_{total} be positive, which requires that

$$H^2 > (8/3)\pi G\rho. \tag{A5}$$

■ The Critical Density

Equation (A5) is what we were looking for. We need only to measure, in the real world, the terms on either side of the above equation and, depending which side is numerically bigger, we know what the fate of the universe will be, whether gravity or the expansion velocity will win. Historically, cosmologists have re-arranged the above equation to define a critical density, ρ_{crit}, such that, if the average density of the universe (ρ) was exactly at this critical level, then the forces of gravity and expansion would be just in balance:

$$\rho_{crit} \equiv H^2 / [(8/3)\, \pi\, G]\, . \tag{A6}$$

Using this definition, (A6), our criterion, (A5), for the fate of the universe now is simply

$$\rho_{crit} > \rho. \tag{A7}$$

The universe will expand forever if its density is less than the critical density given by equation (A7) and, conversely, it will ultimately collapse if its density is greater than the critical density. We put these results in boldface to stress their great importance.

We see the power of mathematical analysis. Our thought experiment and our fanciful constructions of the sphere and the shell have all disappeared. If the matter density measured in the little sphere is found to be either greater than or smaller than this magic value ρ_{crit}, it will tell us all that we need to know. This little imagined piece of the universe will either expand forever or re-collapse, dependent on whether or not at any cosmic time, *t*, its density is more or less than the critical value defined by equation (A6). We have what we sought a simple test to tell us what will be the fate of the universe.

We have done something absolutely astonishing: with simple math and a little thought experiment of the type that Einstein used, we have discovered a fundamental parameter of cosmology that can be measured, and the value of that parameter provides direct information on the evolution of the universe. We must measure the mass in planets, stars, and galaxies, in a representative piece of the universe, divide by the volume of the region studied to obtain the mean density, and we must also measure in that same piece of the universe the rate of expansion, the Hubble constant. Then we compare the mean observed density to the computed critical density to learn the fate of that piece of the universe. If that piece is typical, then we know the fate of the whole.

Of course all of that is easier said than done! Cosmologists found it aesthetically neat to define a new variable (using another Greek symbol, omega), Ω_{matter}, which expressed the same result in dimensionless terms, with Ω_{matter} being the ratio of the actual average density to the critical density:

$$\Omega_{matter} \equiv \rho\,(t)/\rho_{crit}\,(t). \tag{A8}$$

What equation (A8) tells us is that determining the fate of the universe reduces to accurately measuring one dimensionless quantity, Ω_{matter}. (By "dimensionless" number we mean one that does not have any units attached to it like meters or seconds; it is a pure number like 5 or π.) Now, combining equations (A7) and (A8) we learn that we simply must determine if Ω_{matter} is greater than unity, less than unity, or perhaps magically just equal to unity. The task is at once straightforward and awesome.

Later in this appendix we will redo the little example of the expanding sphere with allowance for the cosmological constant—dark energy—and show the dramatic consequences that ensue. But let us look now at what the cosmologists of the 1950s and 1960s attempted, as it was the template for later, more successful efforts. Sandage did not try to determine the fate of the universe by making an accurate inventory of the mass density observed, but rather he tried to make measurements that would tell him, directly, if our little imaginary sphere was decelerating rapidly or slowly.

■ How to Measure the Slowing of the Universal Expansion

To see how in principle a purely geometric test of the cosmological model can be formulated, let us return to the motion of our little test sphere, now using Newton's Second Law and his law of gravity for the motion of the bounding spherical shell, which has a mass that we will write again as m. We will use the notations of high-school calculus in what follows. Force equals mass times acceleration:

$$\begin{aligned} \text{Mass} \times \text{Acceleration} &= m d^2r/dt^2, \\ &= F = -\,\text{G}Mm/r^2, \end{aligned} \tag{A9}$$

or

$$d^2r/dt^2 = -\,\text{G}M/r^2.$$

This is the simple equation for the motion of our little sphere, derived directly from Newton's laws of motion and gravity, and it holds the key to finding the evolution of the universe. The mass of the shell has disappeared from the formula and, as before, we can ignore the rest of the universe in calculating the evolution of the radius of the sphere. We now multiply both sides of equation (A10) by the velocity, $v = dr/dt$:

$$(dr/dt)\,(d^2r/dt^2) = -\,GM(dr/dt)/r^2. \qquad \text{(A10)}$$

Now, both sides of this equation can be written as time derivatives of simple quantities:

$$(1/2)\,(d/dt)\,(dr/dt)^2 = (d/dt)\,GM/r. \qquad \text{(A11)}$$

If we integrate this equation of motion (A11) forward from the present time, t_0, to some future time, we obtain

$$v^2 = (dr/dt)^2 = \text{Constant} + 2GM/r, \qquad \text{(A12)}$$

where

$$\text{Constant} = [v^2 - 2GM/r]_0. \qquad \text{(A13)}$$

This equation tells us how the velocity of expansion of the imaginary shell will change with time. Here, the constant in equation (A12) is to be evaluated at the starting time, when we begin our calculation and the sphere is expanding with just the velocity prescribed by the current Hubble velocity. Looking at equation (A12) we can see without further calculation that, if we want a solution where r can become large on the right-hand side (i.e., for the universe to expand forever), this would make the last term small, and then the constant must be positive, since the left-hand side of the equation is a squared quantity and must always be positive. If the constant is negative, then it is impossible for the radius of the sphere to become large, that is, for the universe to expand forever. Thus the fate of the universe is determined by whether the constant defined by equation (A13) is positive or negative. Now, on rewriting the quantity in the brackets into simpler variables (e.g., $v_0^2 = H_0^2\, r^2$ and $M = (4/3)\,\pi\rho\, r^3$), we see that the condition that allows the radius to become large is simply $v_0^2 > 2GM/r_0$ or

$$H_0^2 > (8/3)\,\pi\, G\, \rho_0. \qquad \text{(A14)}$$

We have exactly recovered the condition that we had derived earlier (inequality A5) for the universe to expand forever; the little sphere must have positive total energy. Reassured by this, let us go back to the original equation (A9) for the acceleration, (d^2r/dt^2) on the right-hand side replace M again by density times volume, multiply it by r, and then divide by $v^2 = (dr/dt)^2$ to obtain another dimensionless "deceleration parameter" which we will name q_0:

$$q_0 \equiv -\,[r\,(d^2r/dt^2)/(dr/dt)^2]_0 = (4\pi/3)\,(G\,\rho_{Sphere}/\,H^2)_0. \quad (A15)$$

Now, by using $H = (dr/dt)/r$ to simplify the middle term, we obtain

$$(d^2r/dt^2) = d/dt\,(Hr) = r\,dH/dt + H^2\,r, \qquad (A16)$$

and, by combining (A15) and (A16), we derive the following neat result:

$$q_0 = -\,[1 + (dH/dt)/H^2] = \tfrac{1}{2}\,\Omega_{matter}. \qquad (A17)$$

We have found something, q_0, defined by equation (A17), that we can measure. It is defined in terms of the Hubble constant, H at the present epoch and the rate of change of the Hubble constant, dH/dt, which is the measurable difference between the Hubble constant now and the Hubble constant at an earlier epoch. If the measured value of q_0 is greater than ½, the universe will re-collapse, but if it is less than ½, the universe expands forever. Sandage simply had to measure the Hubble constant at two redshifts, locally and at a great distance from us (corresponding to an earlier time), to know the answer. Since this new parameter has been mathematically found to be proportional (differing by a factor of ½) to the holy grail, parameter Ω_{matter} (which is so hard to measure) we have now discovered an additional independent observational metric and have derived two methods that the observers at the big telescopes could use to find the correct cosmological model.Another simple thought experiment will give us what physicists call "a sanity check" of this last important equation, to see if it tells us something that we already know to be true. Let us go back to Hubble and suppose that the masses of the galaxies were negligible so that we could neglect gravity. Then every galaxy is flying away from every other galaxy at a constant velocity. Consider any pair, separated by distance D, and moving apart at rate V, then from high school physics $D = Vt$, the separation is just the velocity times the elapsed time since the Big Bang. But we also have Hubble's law (A1), $V = HD$, and combining these two equations and dividing by D we find the wonderfully simple result —$H = 1/t$ that we already noted in chapter 3. This tells us that the Hubble constant would decline with time even without gravity. But of course this is obvious since, if the time doubles, the separation between two galaxies in any pair doubles, but the velocity at which they recede from one another (absent gravity) will not change, and hence the measured Hubble constant will only be half as large at the later time. Now, given that H varies inversely with time, we evaluate the bracket in (A17), the deceleration parameter q_0, and find that the quantities involving H cancel the first term so that the q_0 is zero. Then equation (A17) tells us that Ω_{matter} (the ratio of the actual average density of the universe to the critical density) must also be zero, which of course it would be since we started our thought experiment in a universe that had no gravitating matter. All is well.

As we noted, equations (A15) and (A17) above defining the deceleration parameter tell us that, if our cosmological model is right, then there will be a direct relation between the rate at which the Hubble constant changes with cosmic time and the amount of matter in the universe. We have here what Sandage was looking for in his famous 1961 paper, a quantity, q_0, defined by the middle term of equation (A17), which has a purely geometrical nature and which can be measured by astronomical techniques if the Hubble constant can be observed as a function of time (or distance from us, since the two are related). If we can measure q_0 for any representative piece of the universe, we will have done the equivalent of measuring Ω_{matter}. If $q_0 < (1/2)$ the universe will expand forever.

Thus, we do not need to have a precise inventory of the matter in the universe as we have measured the deceleration directly. The curious reader will immediately ask if direct measurements have verified the neat equation above. We will come back to that question later, but the short answer is "NO." When observational measurements were made in the 1950s and 1960s, the middle term (which defines q_0 in terms of the Hubble constant) was found by Sandage (summarized in 1972) to be consistent with unity or with one-half, but the right-hand side was always found to be very small, on the order of 1 percent.

The measurement uncertainties were sufficiently large that the discrepancy did not overly alarm the community. An insistent reader will now ask, "How is it that this equation passed the 'sanity test' but failed the real test of comparison with observations?" The answer is that we started the thought experiment with the assumption that the only force that could move matter was gravity, and this factual contradiction to our equation must be telling us that something more will be needed. This might have rung the alarm bell and told all sentient cosmologists that perhaps the "something" could be the maligned cosmological constant—dark energy—but, at the time, the reasonable inclination was to let the problem lie, to be reassessed when the observations were more accurate.

Let us return now to the well-behaved world of Newton and look at the details of the standard solution. The equation with which we started this section, for the motion of the test sphere, is solvable by elementary means. Chapter 7, figure 7.2, is a chart illustrating the solutions, showing how the radius of our little test sphere changes in different cosmological models. If Ω_{matter} is greater than unity, or equivalently if q_0 is greater than one-half, the sphere expands for a bit, reaches maximum size, and collapses into a "big crunch," which would surely be fatal for all life in the universe.

If the inequalities are reversed, the universe is "open," and then gravity soon ceases to matter at all and the little spherical shell expands forever at a constant velocity and ever-decreasing density. If one lived inside it and could observe for thousands of billions of years, one would see all of the distant galaxies blink out and the ultimate fate of the observer would be to freeze in an apparently empty universe. Which of these two nasty fates awaits us?

So far, in our mathematical treatment we have taken Newton's laws as sacrosanct. But what if Einstein was correct in introducing the cosmological constant? Then there would be an additional force between any two objects in the universe that was proportional to the distance between them. The extra force had better be so small as to be undetectable on laboratory-length scales and even undetectable on the much larger length scales of the solar system, but still, it might exist and be important on the enormous cosmological scales that we can now explore with our largest telescopes.

So we will go back to our equation of motion (A9), but now add to the right-hand side an extra term, proportional to the radius of our little sphere and thus proportional to this extra cosmological force. Since it derives from the cosmological constant, which uses the Greek letter lambda, we will again use that letter, but now in its uppercase form:

$$d^2r/d\,t^2 = -\,GM/r^2 + \Lambda r/3, \qquad \text{(A18)}$$

where the term with Λ in it represents the new force pushing the galaxies apart. This hypothetical force has the opposite sign to gravity; and also its strength increases with increasing distance, which is opposite to gravity which (more plausibly) decreases in strength with increasing distance. So, given our common experience of physics, this new force is exceedingly strange, but the hypothesis is quite clear and mathematically definite (as represented by equation A18), and so we can test to see if the newly proposed constant of nature, Λ, is—experimentally—positive, negative, or zero.

As before, in order to obtain an equation containing the easier to understand energies, we multiply this equation by the velocity, $v = (dr/dt)$, and integrate it to obtain now

$$v^2 = (dr/dt)^2 = K + 2GM/r + \Lambda r^2/3. \qquad \text{(A19)}$$

Here K is a constant having the value

$$K = [(dr/dt)^2 - 2GM/r - \Lambda r^2/3]_0, \qquad \text{(A20)}$$

which is set by the starting conditions of motion of our little test sphere. Only the last term containing Λ is different from equation (A12), and we have set the "Constant" in (A12) to be K. This new equation, (A19), even though we made only a small change, has a radically different behavior from what we found before, for motion in a universe where there was only gravity and no cosmological constant. On the left-hand side is the square of the expansion velocity of the spherical piece of the universe. On the right-hand side there are three terms—one is a constant and the other two terms have explicit dependencies on the radius of the little sphere that we are considering. If the constant, K, should turn out to be positive (due to a large enough initial velocity, then all the terms on the right-hand side of equation (A19) are positive and the velocity can never go to zero; the expansion can never stop. Then ultimately, as the radius increases, the second gravitational term, which varies inversely with the radius, will become small compared to the new, cosmological constant term which is increasing as the square of the radius and then a totally new (and some might think quite horrid) behavior ensues.

If the radius ever becomes large enough so that the first two terms on the right-hand side of equation (A19) can be ignored and the last term dominates—the new term containing the cosmological constant—then the sphere essentially blows up! It expands exponentially fast, with the solution to (A19) becoming

$$r(t) = \text{const} * \exp[+ t\,(\Lambda/3)^{1/2}]; \qquad \text{(A21)}$$

the little sphere doubles in size again and again and again. It and every part of the whole universe doubles in size each time the age advances by roughly the time equal to the inverse of the Hubble constant. Ultimately, the little sphere and the universe as a whole simply explode in what is called "inflationary behavior." We show in figure 7.2 of chapter 7 a graph of the behavior of our little sphere.

What happens to the Hubble constant in solutions to the equations that contain the cosmological constant? The general case is somewhat complicated, but the most interesting case is easily characterized. That is the "flat" case, in which the constant, K, vanishes and the total energy (including the kinetic energy of expansion, the gravitational energy, and also the dark energy due to the cosmological constant) of the little sphere was initially precisely zero. Then we can take equation (A19) and divide both sides by r^2 and obtain a very simple result. The left-hand side is from equation (A1) just equal to the square of the Hubble constant, H^2. The first term on the right-hand side is by hypothesis zero since $K = 0$. The third term is $2GM/r^3$, which from equation (A4) is just proportional to the density of the sphere, so we have

$$H^2(t) = (dr/dt)^2/r^2 = (8/3)\pi G\rho + \Lambda/3. \tag{A22}$$

Then we use (A6) and (A8) to re-express the matter density in units of the critical density and the parameter Ω_{matter} and finally we define a new parameter to represent the dark energy:

$$\Omega_\Lambda \equiv \Lambda/(3^* H_0^2). \tag{A23}$$

With these definitions equation (A22) becomes

$$H^2(t) = H_0^2[\Omega_{matter,0}(1+z)^3 + \Omega_\Lambda]. \tag{A24}$$

In the past, at large redshift the matter term dominates, but in the future, the dark energy will dominate and we will simply have the Hubble constant to be truly constant in time and Ω_Λ becoming unity. It is clear from equation (A24) that if we could measure the Hubble constant as a function of cosmic time, or equivalently, as a function of redshift or cosmic distance from us, then we could experimentally determine both the matter density now, $\Omega_{matter,0}$, and the energy density locked up in the magical dark energy, Ω_Λ. This is just what the supernova teams did to win the 2011 Nobel Prize in physics. If, now, we take the current epoch to be the moment of interest, then, $z = 1$, $H^2(t) = H_0^2$ and we have the wonderfully simple result:

$$\Omega_{matter}(t) + \Omega_\Lambda(t) = 1. \tag{A25}$$

If a cosmological constant is allowed, this equation provides the new requirement for a flat universe in which the circumference of a circle is always and ever equal to π times the diameter of the circle. The matter density and the energy density, in cosmological units, should always, at all times add up to exactly unity, and this seems to be the case, experimentally, to within several percent accuracy.

Appendix Two

How Do We Measure Mass in Astronomy

We will start by using the Moon's orbit to find the mass of the Earth. This we do just as a check of the method we will be using for finding the masses of other astronomical bodies, since we have already measured the Earth's mass using laboratory methods.

We write M_m as the mass of the Moon and R_m as the distance from the center of the Earth to the center of the Moon. Substituting these into the equation for gravitational attraction, we get the value of the attracting force of gravity acting between the Earth and the Moon:

$$F_g = GM_mM_e/R_m^2. \quad \text{(B1)}$$

As Newton realized, this must be exactly balanced by the centripetal force pushing the Moon away from the Earth. If the Moon's period (one month) is P_m, then its velocity in its orbit (taken for simplicity here to be circular) is found by dividing the circumference of the orbit by the period. That is to say,

$$v = 2\pi R_m/P_m. \quad \text{(B2)}$$

Then the centripetal force is

$$F_c = v^2M_m/R_m = 4\pi^2R_mM_m/P_m^2. \quad \text{(B3)}$$

Now, since the force of gravity and the centripetal force must balance, $F_g = F_c$, we have from (B1) and (B3)

$$M_e = 4\,\pi^2 R_m{}^3/(G\,P^2{}_m). \qquad \text{(B4)}$$

Notice that the unknown mass of the Moon has dropped out of the equation, and we have all known astronomical quantities on the right-hand side. Physically, the fact that our test body, the Moon, has disappeared from our equation is due to the same wonderful fact that Galileo discovered and that Einstein made the cornerstone of general relativity; all bodies fall at the same rate in the same gravitational field, so the motion of the Moon is independent of its mass.

Putting into these equations the period of the Moon's orbit, one month $P_m = 2.4 \times 10^6$ seconds, the distance of the Moon from the center of the Earth, $R_m = 3.8 \times 10^8$ meters, and G from our earlier calculation, we obtain the mass that the Earth must have to keep the Moon in its orbit: 6.0×10^{21} kg. It agrees very well with what is obtained by the Cavendish experiment. We used here the distance of the Moon from the Earth, but this could be (and actually was, by Greek astronomers) determined by elementary geometry; the edge of the Moon, seen at a given time from two different places on Earth, is seen at a slightly different spot in the sky. Then, given the number of miles between the two observers and the angle between the two spots on the sky, the distance to the Moon, in miles, can be computed by elementary trigonometry.

The method outlined above is the fundamental means for almost all determinations of the masses of astronomical bodies. As a first application, we can find the mass of the Sun. The distance between the Earth and the Sun is $R_e = 1.5 \times 10^8$ km, and for the period we use the year, $P_e = 3.15 \times 10^{7}$s. Plugging these values into equation (B4) we obtain $M_{sun} = 2.0 \times 10^{30}$ kg.

Now let us check if this method is giving a sensible answer. Instead of using the Earth's orbit, we could take that of Venus, which is closer to the Sun than we are, or the orbit of Mars, which is farther out, and again compute the mass that the Sun would need to keep these planets steadily circling in their orbits. When we do this, we obtain the identical mass for the Sun.

Since our Sun circles around the center of our own spiral galaxy, the Milky Way, we can again use equation (B4) to obtain its mass interior to our orbit. Using for the radius in (B4) the distance from us to the center of our galaxy, 8.5kpc, and for the period, 240 million years, we find a mass equivalent to 90 billion suns inside of our orbit. While this is a large number, it is not unreasonably large compared to the counted number of stars in the southern Milky Way, if we allow for the fact that many are obscured by dust between us and them. Almost all methods of measuring mass in astronomy are variants of these simple calculations.

Glossary

anisotropy Irregularities on a sky map which may indicate cosmic structures.

anthropic principle A philosophical construct based on the requirement that the physical parameters of our universe must be compatible with the existence of intelligent observers.

apparent brightness The brightness of an object as seen from Earth.

apparent magnitude The standard measure for the apparent brightness of astronomical objects; the scale was originally set by Arab and central Asian astronomers. Smaller numbers designate brighter objects.

baryon acoustic oscillations The sound waves propagating in the early universe are like the waves coming out of an organ pipe, and measuring their properties on the sky can determine cosmic parameters in the same way that we can tell the length of the organ pipe by the depth of the sound.

baryonic matter Ordinary matter containing protons and neutrons and the ordinary chemical elements; not dark matter.

beta decay A type of radioactive decay in which an atom emits a beta particle (e.g., an electron).

big bang model The theory that describes how the universe was formed 13.7 billion years ago in an explosion from an extraordinarily dense initial state. The hot big bang model, now standard, also presumes that the early state of the universe was extraordinarily hot, and the observed cosmic background radiation shows the ashes of this primeval fireball.

Big Crunch The opposite of the Big Bang, in which the universe would end if gravity were strong enough to reverse the expansion of space and cause a re-collapse.

black hole An object that has collapsed to such a small volume that the gravitational field created does not allow light to escape. Supermassive black holes (from a million to many billion times more massive than the Sun) reside at the center of most massive galaxies.

boundary conditions In solving equations, the conditions that apply at the boundary between the volume studied and the rest of the world, so the boundary condition for water sloshing in a bottle is that the water never leaves the bottle.

Cosmic background radiation (CBR) The Cosmic Background Radiation field is the radiation at our cosmic epoch seen primarily as radio waves that is the remnant of the very hot radiation field left over from the hot big bang explosion.

CCD Charge-coupled device, a detector similar to what is in digital cameras that, when used in astronomy, greatly increases the power of telescopes to see distant objects.

celestial sphere The imaginary sphere on which objects in the sky appear to move while the Earth rotates.

Cepheids A special kind of variable star that helps astronomers measure cosmic distances because of the precise relation between its period of oscillation and its intrinsic luminosity.

closed universe If the density of the universe is greater than the critical density, the geometry of space-time is such that a circle has a circumference less than π times its diameter and, as on a sphere, one can travel in one direction but come around to where one started.

CNO cycle The reactions that occur in high-mass stars that fuse hydrogen into helium using carbon and nitrogen as catalysts in the reactions.

Cosmic Background Explorer (COBE) The satellite launched in 1989 that detected the fluctuations in the cosmic background radiation field which reflected the primordial perturbations that had seeded the growth of galaxies and all cosmic structure.

cosmic rays Atomic nuclei and various subatomic particles that travel through space at nearly the speed of light.

cosmological constant The term in Einstein's field equations of general relativity that represents a repulsive force or energy (dark energy) that pushes matter from other matter and causes the expansion of the universe to accelerate.

cosmology The study of the nature, overall structure, and evolution of the universe.

covariant equations Equations written in a special form so that they will be true for all observers regardless of the velocity at which they are moving.

critical mass density The precise mass density that lies between a universe that re-collapses and one that expands forever.

cyclic model A current version by Steinhardt and Turok of a model universe that could have repeated collapse and expansion phases.

dark ages The period after the radiation and gas from the Big Bang had cooled down due to expansion and before stars and galaxies had begun to form.

dark energy Mysterious energy that is a property of space that causes the expansion of the universe to accelerate.

dark matter Mysterious matter that dominates the total mass of the universe. It neither emits nor absorbs light, but we infer its existence from the gravitational forces it exerts on objects that we can see, and it responds to gravitational forces in the same fashion as normal matter.

deceleration parameter A measurable, dimensionless number that expresses the rate at which gravity is causing the expansion of the universe to slow.

dipole variation The cosmic background radiation is slightly brighter in the octant of the sky toward which we are moving and slightly fainter in the opposite one.

disk The flat, more or less circular portion of a spiral galaxy that contains the spiral arms, that will look elongated if seen edge on.

Doppler shift The shift in wavelength and frequency of photons emitted from an object that is moving toward or away from the observer.

electromagnetism One of the four fundamental forces; it governs interactions between atoms and molecules.

elliptical galaxy Smooth galaxies with a rounded distribution of stars that typically contain orange-red, older star populations as compared to spiral galaxies.

energy density The average energy per unit volume in the entire universe.

entropy The degree of disorder in a physical system.

epicycle Epicycles were the extra "wheels on wheels" that astronomers in antiquity needed to explain the apparent motions of the planets on the sky before the sun-centered solar system was understood.

epoch of reionization Era in the early universe when the first stars were formed.

era of recombination Time in the early universe when the hot soup of protons and electrons cooled enough to combine into atoms of hydrogen.

escape velocity The velocity an object would have to achieve to escape from the gravitational field of a massive body.

Euclidean geometry The geometry that we learn in high school that is valid on a flat plane or piece of paper.

extragalactic Outside of our Milky Way galaxy.

filaments Large, cosmic structures that consist of numerous individual galaxies in a gravitationally connected linear structure.

flat universe The geometry of space-time is flat and a circle has a circumference exactly π times its diameter as it has on a plane. On a sphere the circumference is less than π times its diameter, and on a saddle it is more. In a universe without a cosmological constant, the flat universe has exactly the critical density.

flux Another term for the apparent brightness of an object, the unit usually being energy passing through a square centimeter per second.

general theory of relativity Einstein's theory (1915) that extends special relativity to include cases of accelerating observers. Rather than thinking of gravity as a force, Einstein proposed that gravity is the curvature of space-time.

globular cluster Round stellar system containing roughly one hundred thousand stars each. They reside primarily in the spherical component surrounding galaxies and contain very old stars.

Goldilocks problem The situation where cosmic parameters seem to have been finely tuned to be "just right" to allow for the universe that we see today.

Grand unified theory (GUT) A concept that three of the prime forces of physics (electromagnetism, the strong force, and the weak force) can be combined into a single theory at sufficiently high energies.

gravitational lens An object creating a significant gravitational field that causes light to bend. The image we detect will be magnified and often distorted.

gravitational redshift The spectral shift of light toward the red caused by its losing energy as it climbs out of a gravitational field as a photon moves toward an observer.

great wall A large sheet-like structure composed of numerous galaxies.

halo The roughly spheroidal region enveloping a galaxy. While it may contain stars, most of its mass is in the form of dark matter.

heat death The idea that a universe that lasted forever would ultimately run out of fuel and everything would cool down to a very low temperature.

Higgs boson Hypothetical elementary particle that naturally should exist in the standard model of particle physics and whose tentative discovery was announced by the Large Hadron Collider team on July 4, 2012.

Hubble constant A number, H_0, that represents the current rate at which the universe is expanding; when multiplied by the distance to the object in mega-parsecs it gives its velocity of recession in km/s.

Hubble law The law which states that the farther away a galaxy is from the Milky Way, the faster it moves away from us, the velocity being proportional to the distance.

hydrostatic equilibrium Describes a state in which the inward gravitational forces are exactly balanced by the outward pressure forces, that, in turn, arise from the hot interior of a star.

inflation The idea that the universe had a period of exponential growth in its earliest phases.

interstellar medium The gas and dust that are found between the stars.

intrinsic luminosity The total amount of energy emitted from a star each second.

inverse square law (gravitation) The law, due to Isaac Newton, which states that the gravitational attraction between two objects is proportional to their masses and inversely proportional to the square of the distance between them.

ionization The process by which electrons are stripped from an atom.

irregular galaxy Galaxies whose shapes are neither especially elliptical nor spiral.

island universe A term coined by Immanuel Kant to refer to the idea that the spiral nebulae are galaxies similar to our own Milky Way.

isotopes Atoms of the same element that differ in the number of neutrons they contain. Different isotopes of the same element weigh different amounts, and can have different nuclear reactions, but they have essentially the same chemical properties.

isotropic The same in all directions. Looking around on a very foggy day, the brightness of the light in all directions is isotropic.

Jeans instability An explanation of gravitational instabilities (discovered by James Jeans) that cause massive structures to grow from very small initial fluctuations.

Jeans mass The critical mass for which an interstellar cloud will begin a runaway gravitational collapse.

LCDM Abbreviation for Lambda-Cold Dark Matter, which is the standard model of big bang cosmology that incorporates both dark matter and dark energy.

LHC Large Hadron Collider, the world's largest and highest energy elementary particle accelerator, first activated in 2008 with hopes to find the Higgs particle and also perhaps to detect dark matter particles.

light-year The distance a beam of light (moving at a speed of 300,000 km/s) can travel in one year, equal to 9.5×10^{12} km.

local group The group of galaxies that contains our own Milky Way galaxy, Andromeda (M31), M33, and numerous smaller systems.

mass density The average mass per unit volume of all the matter in the universe.

mass-to-light ratio The mass of an object divided by its luminosity usually given in solar units.

Messier catalog A catalog of extended objects in the sky studied by Charles Messier (published in 1771) and denoted by a capital M followed by a number, e.g., M81.

microwave background radiation The remnant radiation that radio telescopes detect from a time just 380,000 years after the Big Bang, when the universe first became transparent.

molecular clouds Dense clouds of gas in interstellar space that are made up primarily of hydrogen molecules. Star formation occurs in molecular clouds.

nebula Presently defined as an extended cloud of dust and gas that is often seen glowing from absorbed or reflected starlight. Before galaxies were understood as comprising numerous stars too close together to be identified individually, they were often called nebulae.

nebular origin of the solar system The theory that describes how our solar system formed out of a rotating cloud of dust and gas over 4 billion years ago.

neutrino A very low-mass elementary particle.

neutron star A type of star, approximately the mass of the Sun, but only about 10 km in size, made almost entirely of neutrons as dense as the nucleus of an atom. Probably made in supernova explosions.

nuclear fission The process in which a large atomic nucleus splits into smaller nuclei with enormous release of energy, as in an atomic bomb.

nuclear fusion The process by which small atomic nuclei are joined together to create a larger nucleus.

open universe If the density of the universe is less than the critical density, the geometry of space-time is such that a circle has a circumference greater than π times its diameter. An open universe can never re-collapse.

parallax The apparent shift in an object's position in the sky due to viewing it from different positions.

parsec (pc) The distance an observer would be from our solar system to see our orbit around the Sun as being one arc second in angle on the sky. One parsec is several times farther than the distance light can travel in a year and is roughly equal to 30 trillion kilometers.

perihelion The point in an object's orbit at which it is closest to the Sun.

period The time it takes for a variable star to complete one full cycle in luminosity.

perturbation A small change in a system or part of a system caused by an external force such as gravity.

phase transition The abrupt transition from one state of matter to another, e.g., the transition from water to ice.

photon An individual particle of light.

Planck era The very early era when the universe was younger than 5.4×10^{-44} seconds, and our current theories of gravity and quantum mechanics come into conflict.

plasma An ionized gas in which the electrons have been separated from the atoms to which they were bound.

population I stars Young, typically bright blue stars in the spiral arms of galaxies.

population II stars Older stars in the central region of spiral galaxies and globular clusters typically fainter and redder than population I stars.

precession The gradual wobble than an object experiences as it rotates.

principle of relativity The idea that the fundamental laws of physics must be the same for all observers moving at a constant velocity with respect to each other.

proton-proton cycle The reactions that occur in low-mass stars that fuse hydrogen into helium.

pulsar A neutron star that rotates rapidly such that we detect periodic pulses of radiation from it, as when a beam from a lighthouse passes by us.

quantum gravity At attempt to unify quantum mechanics and gravity into a single model.

quantum mechanics A description of physical phenomena at the smallest scales where discreteness and randomness are essential properties. It describes the interactions of atoms, subatomic particles, and photons.

quasars The especially luminous centers of some galaxies that are powered by mass from the galaxy falling into the supermassive black hole at the galactic center.

quintessence A possible force field that causes the expansion of the universe to accelerate: a variety of dark energy.

radiometer Instrument for measuring the strength of radio signals.

relativity *See* special theory of relativity, general theory of relativity

rotation curve The change in the rate of rotation of orbiting bodies with increasing distance from the center of the system. In the solar system, the outer planets rotate around the Sun at a slower rate than the inner ones.

SDSS Sloan Digital Sky Survey was an automated survey of the northern sky (starting in 2000) using CCDs to catalog and measure the properties of one million galaxies and one hundred thousand quasars, the most comprehensive astronomical database ever assembled.

space-time continuum The mathematical combination of the three dimensions of space and the one dimension of time.

special theory of relativity Einstein's theory (1905) that describes the idea that all motion is relative and that the speed of light in empty space is the same for all observers in non-accelerating frames of reference.

spectral lines Bright or dark lines that appear in graphs of an object's distribution of photons at various wavelengths or frequencies. By studying these lines, astronomers can identify the chemical composition of the objects they see and the velocity of the object with respect to the observer.

spectroscope A prism-like device that spreads out light like a rainbow into its different colors.

spiral galaxy A galaxy composed of a large, flat disk of stars, often with a bulge at the center. Spiral arms within the disk are filled with younger, blue star-forming regions.

standard candle An object that is useful for measuring cosmic distances because its true luminosity is known.

static universe The model universe proposed by Einstein in 1917 in which gravity is exactly balanced by the cosmological constant and nothing moves. It was proposed before Hubble's results were widely known, and is unstable.

steady state model A theory that held that the universe is eternal, having neither a beginning nor an end, appearing the same at all times and places. Matter must be created everywhere continuously in this cosmological model.

string theory An attempt to unify quantum mechanics and general relativity into a single model.

strong force One of the four fundamental forces; it holds the nucleus together by overcoming electrostatic repulsion.

supernova A star that explodes at the end of its lifetime, either totally disrupting or leaving a condensed object like a black hole or neutron star.

tensor In general relativity, a four-by-four array of numbers used to describe quantities in space-time.

thermal blackbody radiation The spectrum of radiation produced by an ideal, opaque object of a definite temperature. The hotter the object, the more the light is shifted from the red to the blue or ultraviolet.

Type Ia and Type II supernovae The first is a special kind of supernova that always emits nearly identical luminosities when it explodes, thus allowing astronomers to use it as a standard candle to measure cosmic distances. They arise from relatively low mass stars. The second is from the explosion of massive stars, and they are quite variable in their brightness.

universal gravitation Isaac Newton's realization that all clumps of matter, from atoms to clusters of galaxies, attract each other.

variable star A star whose brightness oscillates in a regular way.

vector An array of three numbers showing both quantity and direction. Velocity is a vector.

virial theorem A statistical method for relating the total kinetic energy of a system of orbiting bodies to its total potential energy if it is to stay in equilibrium and neither collapse nor expand.

weak force One of the four fundamental forces; it is responsible for radioactive decay.

weakly interacting massive particle (WIMP) A popular candidate for a dark matter elementary particle.

Wilkinson Microwave Anisotropy Probe WMAP, launched in 2001, that has provided the most precise analysis of the cosmic background radiation and has measured the cosmic parameters with sufficient precision to confirm the LCDM standard cosmic model.

Bibliography

Bartusiak, Marcia. *Through a universe darkly: a cosmic tale of ancient ethers, dark matter, and the fate of the universe.* New York: HarperCollins, 1993.

Bartusiak, Marcia. *Archives of the universe: a treasury of astronomy's historic works of discovery.* New York: Pantheon Books, 2004.

Berger, André. *The Big Bang and Georges Lemaître: proceedings of a symposium in honour of G. Lemaître fifty years after his initiation of Big-Bang Cosmology, Louvain-la-Neuve, Belgium, 10–13 October 1983.* Dordrecht: D. Reidel, 1984.

Bertotti, B., R. Balbinot, S. Bergia, and A. Messina. *Modern cosmology in retrospect.* Cambridge: Cambridge University Press, 1990.

Bondi, Hermann. *Cosmology.* 2nd ed. Cambridge: Cambridge University Press, 1960.

Bondi, Hermann. *Rival theories of cosmology; a symposium and discussion of modern theories of the structure of the universe.* London: Oxford University Press, 1960.

Christianson, Gale E. *Edwin Hubble: mariner of the nebulae.* New York: Farrar, Straus, Giroux, 1995.

Clerke, Agnes M. *A popular history of astronomy during the nineteenth century,* 4th ed. London: A. and C. Black, 1902.

Eddington, A. S. *Stars and atoms.* Oxford: Clarendon Press, 1927.

Einstein, Albert. *The meaning of relativity, with a new introduction by Brian Greene.* Princeton: Princeton University Press, 2005.

Farrell, John. *The day without yesterday: Lemaître, Einstein, and the birth of modern cosmology*. New York: Thunder's Mouth Press, 2005.

Gamow, George. *The creation of the universe*. New York: Viking Press, 1961.

Gingerich, Owen. *The book nobody read: chasing the revolutions of Nicolaus Copernicus*. New York: Walker & Company, 2004.

Gold, Thomas. *Taking the back off the watch: A personal memoir*. Heidelberg: Springer, 2012.

Guth, Alan H. *The inflationary universe: the quest for a new theory of cosmic origins*. Reading, MA: Addison-Wesley, 1997.

Harrison, Edward Robert. *Cosmology: the science of the universe*. 2nd ed. Cambridge: Cambridge University Press, 2000.

Harrison, Edward Robert. *Masks of the universe: changing ideas on the nature of the cosmos*. 2nd ed. Cambridge: Cambridge University Press, 2003.

Hoskin, Michael A. *The Cambridge illustrated history of astronomy*. Cambridge: Cambridge University Press, 1997.

Hoskin, Michael A. *Discoverers of the universe: William and Caroline Herschel*. Princeton: Princeton University Press, 2011.

Hubble, Edwin Powell. *The realm of the nebulae*. New Haven: Yale University Press, 1936.

Kirshner, Robert P. *The extravagant universe: exploding stars, dark energy, and the accelerating cosmos*. Princeton: Princeton University Press, 2002.

Kragh, Helge. *Cosmology and controversy: the historical development of two theories of the universe*. Princeton: Princeton University Press, 1999.

Laplace, Pierre-Simon. *Exposition du dystem du monde*. Paris: Imprimerie du Cercle-Social, 1796.

Longair, Malcolm S. *The cosmic century: a history of astrophysics and cosmology*. Cambridge: Cambridge University Press, 2006.

Millar, Ian. *Chambers biographical dictionary of scientists*. Edinburgh: Chambers, 1989.

Mitton, Jacqueline. *Cambridge illustrated dictionary of astronomy*. Cambridge: Cambridge University Press, 2007.

Mitton, Simon. *Exploring the galaxies*. New York: Scribner, 1976.

Mitton, Simon. *Fred Hoyle: a life in science*. Pbk. ed. Cambridge: Cambridge University Press, 2011.

Murdin, Paul. *Secrets of the universe: how we discovered the cosmos*. Chicago: University of Chicago Press, 2009.

Naselsky, Pavel D., and Dmitry I. Novikov. *The physics of the cosmic microwave background*. Cambridge: Cambridge University Press, 2006.

Nussbaumer, Harry, and Lydia Bieri. *Discovering the expanding universe*. Cambridge: Cambridge University Press, 2009.

Osterbrock, Donald E. *James E. Keeler: pioneer American astrophysicist*. Cambridge: Cambridge University Press, 1984.

Osterbrock, Donald E. *Walter Baade, a life in astrophysics*. Princeton: Princeton University Press, 2001.

Peebles, P.J.E. *Principles of physical cosmology*. Princeton: Princeton University Press, 1993.

Peebles, P.J.E., L. A. Page, and R. B. Partridge. *Finding the big bang*. New York: Cambridge University Press, 2009.

Rees, Martin J. *Just six numbers: the deep forces that shape the universe*. London: Weidenfeld & Nicolson, 1999.

Roberts, Isaac. *A selection of photographs of stars, star-clusters and nebulae*. London: Universal Press, 1893.

Ryle, Martin, Nicholas Kurti, and Robert Boyd. *Search and research*. London: Mullard Ltd, 1971.

Sciama, D. W. *Modern cosmology and the dark matter problem*. Cambridge: Cambridge University Press, 1993.

Shapley, Harlow. *The inner metagalaxy*. London: Oxford University Press, 1957.

Silk, Joseph. *The big bang*. Rev. and updated ed. New York: W. H. Freeman, 1989.

Steinhardt, Paul J., and Neil Turok. *Endless universe: beyond the Big Bang*. 1st ed. New York: Doubleday, 2007.

Thoren, Victor E., and J. R. Christianson. *The Lord of Uraniborg: a biography of Tycho Brahe*. Cambridge: Cambridge University Press, 1990.

Weinberg, Steven. *The first three minutes: a modern view of the origin of the universe*. New York: Basic Books, 1993.

Index